Key to the Diversity and History of Life

Alexey Shipunov

December 24, 2019

Shipunov, Alexey. *Key to the diversity and history of Life*
December 24, 2019 version. 114 pp.
URL: http://ashipunov.info/shipunov/school/biol_111/ph_key/ph_key.pdf

This book is for any biology student who wants to know the subject better. I concentrate on the history of life, diversity of plants, protists and animals, and some other general aspects including elements of biogeography.

On the cover: Nikolaj Zinovjev, "Devonian Period" (1968)

This book is dedicated to the public domain

Contents

Chapter 1

The Really Short History of Life

History of Earth is split in multiple intervals, and some of them are listed in the clock and chart below. These classifications, however, do not reflect well the stages of evolution. This is why in this chapter, the history of life is described from the *palaeoecological* point of view which reflects milestones of organic world development.

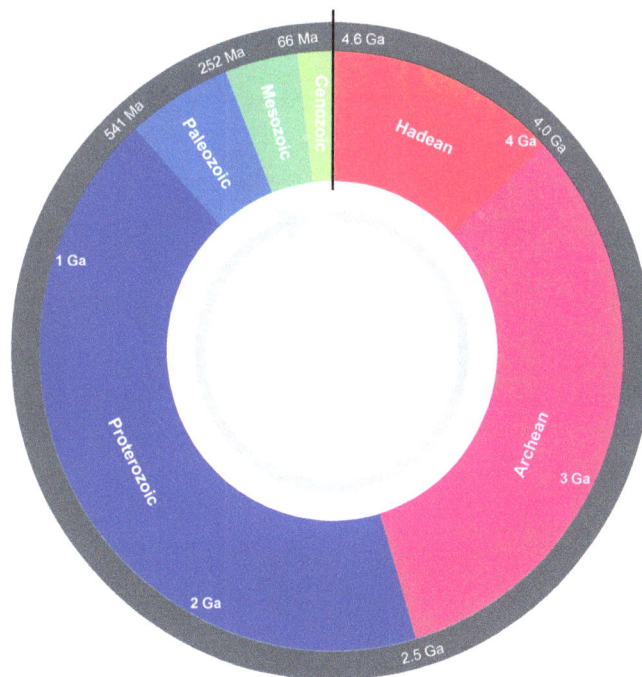

Geologic clock.

Chart 1 — Phanerozoic (Cenozoic / Mesozoic)

Series / Epoch	Stage / Age	numerical age (Mya)
Holocene		present / 0.0117
Pleistocene (Quaternary)		2.58
Pliocene (Neogene)		
Miocene (Neogene)		23.03
Oligocene (Paleogene)		
Eocene (Paleogene)		
Paleocene (Paleogene)		61.6
	Danian	66.0
Upper (Cretaceous)	Maastrichtian	72.1 ±0.2
	Campanian	83.6 ±0.2
	Santonian	86.3 ±0.5
	Coniacian	89.8 ±0.3
	Turonian	93.9
	Cenomanian	100.5
Lower (Cretaceous)	Albian	~113.0
	Aptian	~125.0
	Barremian	~129.4
	Hauterivian	~132.9
	Valanginian	~139.8
	Berriasian	~145.0

Eonothem: Phanerozoic. Erathem: Cenozoic, Mesozoic.

Chart 2 — Phanerozoic (Mesozoic / Paleozoic)

Series / Epoch	numerical age (Mya)
Upper (Jurassic)	~145.0
Middle (Jurassic)	
Lower (Jurassic)	
Upper (Triassic)	201.3 ±0.2
Middle (Triassic)	
Lower (Triassic)	251.902 ±0.024
Permian	298.9 ±0.15
Pennsylvanian (Carboniferous)	
Mississippian (Carboniferous)	358.9 ±0.4

Chart 3 — Phanerozoic (Paleozoic)

Series / Epoch	numerical age (Mya)
Upper (Devonian)	358.9 ± 0.4
Middle (Devonian)	
Lower (Devonian)	419.2 ±3.2
Silurian	443.8 ±1.5
Upper (Ordovician)	
Middle (Ordovician)	
Lower (Ordovician)	485.4 ±1.9
Cambrian	541.0 ±1.0

Chart 4 — Precambrian

Eonothem / Eon	Erathem / Era	System / Period	numerical age (Mya)
Precambrian	Proterozoic	Neo-proterozoic — Ediacaran	541.0 ±1.0 / ~635
		Cryogenian	~720
		Tonian	1000
		Meso-proterozoic	
			1600
		Paleo-proterozoic	
			2500
	Archean	Neo-archean	
		Meso-archean	
		Paleo-archean	
		Eo-archean	4000
	Hadean		~4600

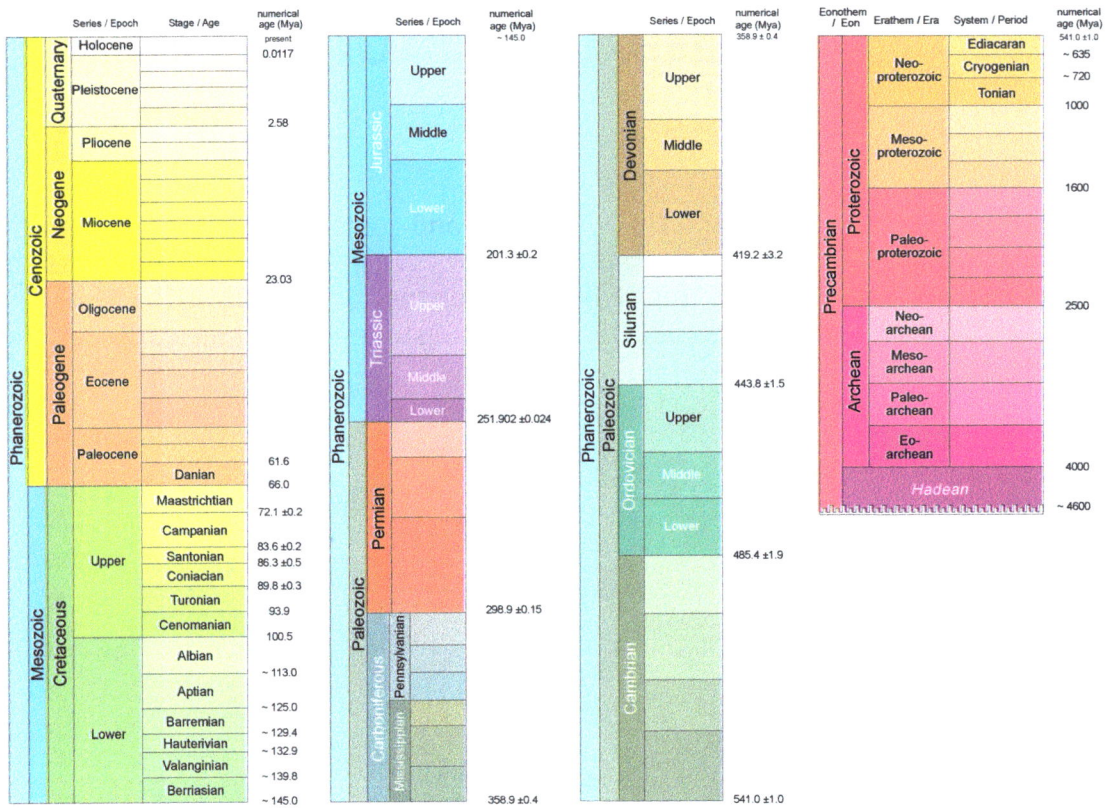

Chronostratigraphic chart. Simplified from http://www.stratigraphy.org/index.php/ics-chart-timescale.

1.1 Origin of Life

Nikolaj Zinovjev, "Archeozoic Era" (1968)

In the strict sense, origin of life does not belong to biology. In addition, biologists were long fought for the impossibility of a spontaneous generation of life (which was a common belief from Medieval times to the end of 19 century). One of the founders of genetics, Timofeev-Resovsky, when he was asked about his point of view on the origin of life, often joked that "he was too small these times, and do not remember anything".

However, contemporary biology can guess something about these times. Of course, such guesses are no more than theories based on common scientific principles, **actuality** and **parsimony**.

First, Earth was very different. For example, the atmosphere had no oxygen; it was much closer to the atmosphere of Venus than to the atmosphere of contemporary Earth and contained numerous chemicals which are now poisonous for most life (like CS_2 or HCN). However, by the end of Archean first oxygen appears in the atmosphere, and in early Proterozoic, it started to accumulate rapidly. The process is called the "oxygen revolution," and it had many consequences. But what the reason for oxygenation was nothing else than the appearance of first photosynthetic organisms, most likely **cyanobacteria**.

Second, the first traces of life on Earth are suspiciously close to the time of Earth origin (4,540 mya)— **molecular clock** place **LUCA** about 4,000 mya, and recently found first traces of cyanobacteria are 3,700 mya. Altogether, life on Earth was most of the time of its existence!

Third, first living things were most likely **prokaryotes** (Monera, bacteria). These could be both photosynthetic (cyanobacteria) and **chemotrophic** bacteria, as evidenced from isotope analysis of Isua sedimentary rocks in Greenland, and now also from the presence of **stromatolites**, the traces of cyanobacteria in the same place.

What was the first living thing? It has a name LUCA, Last Universal Common Ancestor, but only a little could be estimated about its other features. It was probably a cell with DNA/RNA/proteins stream, like all current living things. Unclear is how this stream appeared and how it happened that it was embedded into the cell. One of the helpful ideas is "RNA world", speculation about times when no **DNA** yet exist, and even **proteins** did not function properly, but **RNAs** already worked as an information source as well as biological machines. Another possibility is that **lipid** globules, some other organic molecular and water formed **coacervates**, small droplets in which these RNAs could dwell. If this happened, then resulted structure could be called "proto-cell".

1.2 Prokaryotic World

Most of the Proterozoic prokaryotes (Monera) dominated the living world. Typical landscape these times was high, almost vertical rocks and shallow plains, which should be covered with the tide for dozens of kilometers. This is because there were no terrestrial organisms decreasing erosion. Ocean was low oxygenated; only water surface contained oxygen.

In those conditions, ancestor of eukaryotes appeared. First eukaryotes could probably remain contemporary heterotrophic Excavata (Fig. 2.3) like *Jacoba*, but there are no fossils of this kind. However, there is a number of fossils which could be treated as **algae**, photosynthetic protists. These fossils remind contemporary red and green algae (Fig 2.9, the bottom row). It is possible that some other Proterozoic fossils (acritarchs) belong to other protist groups, for example, unicellular Dinozoa (Fig. 2.8).

Ecosystems of these times were similar to Archean and mostly consisted of cyano- and other bacteria, and represented now by stromatolites. No one can say anything about terrestrial life in Proterozoic, but it possible that Monera dominated there as well.

At the end of middle Neoproproterozoic, continents of Earth joined in one big continent Rodinia; this triggered the most powerful glaciation in history, "snowball Earth", Cryogenian glaciation.

1.3 The Rise of Nonskeletal Fauna

This mentioned above glaciation possibly, in turn, triggered the evolution of Earth, because, in the Ediacaran period (the last period of Proterozoic), animals and other multi-cellular organism appear. There are three most unusual things about Ediacarian ecosystems. First, they were filled with creatures as similar to contemporary life as would (not yet discovered) extra-terrestrial life be. In other words, they (like *Pteridinium*, see Fig. 2.30) had no similarity with the recent fauna and flora. Second, all these Ediacaran creatures were soft, nonskeletal. This last fact is even more striking because, in the next period (Cambrian), almost all animals and even algae had skeletal parts.

There were different types of ecosystems in Neoproterozoic. However, in essence, they all consisted of these soft creatures (it is not easy to say what they were, animals, plants of colonial protists). They thrived for about 90 million years and then suddenly declined (some left-overs existed in Cambrian, though). This decline is the third bizarre thing. Weird because later ecosystems almost always left descendants, even famous dinosaurs went extinct but left the great group of birds, their direct "offspring".

Why they went extinct, it is not clear. Several factors could be blamed: oxidization of ocean, the appearance of macroscopic carnivores, increased transparency of water. The last could relate with two first by means of *pellet production*. Many recent small plankton invertebrates pack their feces in granules (pellets), which speedily fall to the ocean bottom. In Ediacaran, there was probably no pellet production, and therefore ocean water was mostly muddy. When first pellet producers appear, water start to be increasingly more transparent, which raised oxygen production by algae and, as the next step, allowed more and bigger animals to exist. Bigger plankton animals mean that it starts to be rewarding to hunt them (remember ecological pyramid). These hunters were probably first macroscopic carnivores, which caused the end of Ediacaran's "soft life".

After Ediacaran great extinction (this is the first documented great extinction), one can observe the rise of very different creatures, small, **skeletal** Cambrian organisms.

They appear insignificant diversity and represent many current phyla of animals. This is called "Cambrian revolution", or "Cambrian explosion" (see below).

1.4 Filling Marine Ecosystems

Nikolaj Zinovjev, "Silurian Period" (1968)

This happened during the Cambrian and Ordovician periods, which jointly continued for almost 100 million years. Most of this time the Earth climate was relatively warm, but continents were concentrated in the Southern hemisphere. At the end of Ordovician, Africa hit the South Pole, and this resulted in a serious glaciation.

The sea, in large degree, prevailed over the land and thus created exceptionally favorable conditions for the development of marine communities, which in this epoch became finally similar to what we see around now. For some groups, there was not "enough space" in the sea, and, as a consequence, the *colonization of land* from higher organisms started.

At this time, all main types and even classes of invertebrates and vertebrates and terrestrial plants already existed. Stromatolites went to the "background" of ecosystems and were replaced with other builders of bioherms (reef-like organic structures) like archaeocyaths (Fig. 2.13, group probably close to the sponges) and calcareous red and green algae. Archaeocyaths went extinct at the end of the Ordovician, but calcareous algae have survived.

In Cambrian, there was a great variety of different groups of animals, usually *small size and with a skeleton* of different types (phosphate, calcareous, organic): that was a consequence of "skeletal revolution". some of them were crawlers, some swimmers, and some burrowers.

Among the seafloor bilaterians, trilobites (an extinct group of arthropods) dominated, there were also many other groups of arthropods and lobopods (intermediates between ecdysozoan nematode-like "worms" and arthropods), plus various spiralians, namely brachiopods and mollusks (Fig. 2.21, 2.22) including cephalopods which played the role of pelagic predators, preceding sea scorpions and armored fish. There were also plenty of echinoderms, mostly sea lilies and many other, now extinct, classes (Fig. 2.15). First jawless fishes (Fig. 2.17, top row) were also the part of pelagic life.

It can be assumed that at this time started the mass "exodus" of invertebrates to the land. Perhaps, there was already some soil fauna, consisting of nematodes, small arthropods, and other similar organisms.

Green algae were gradually replaced red algae in communities. For some of them, like for some invertebrates, there was "not enough space" in the ocean, and they proceeded to conquer the land. The living conditions outside of the ocean were much more stringent for plants than for the animals, so the process of adaptation took a long time. The first land plants are known from the Ordovician; they probably were liverworts (Fig. 2.10, top left). Land conquest for plants was concerted with the development of symbiosis with mycorrhizal fungi (Fig. 2.5). Apparently, among the first terrestrial photosynthetic organisms were symbioses, both with a predominant fungus and predominant alga. The first gave rise to the lichens, who took the most extreme habitats, and the second to the contemporary terrestrial plants.

Terrestrial plants had to solve many problems. There were, in particular, water supply (so they developed vascular system), gas exchange (acquired stomata), competition for light (body began to grow vertically with the help of supportive tissues), and spore dispersal (diploid stage, sporophyte, began to form sporangia on a long stalk containing spores covered with thick envelope).

A serious plant problem was also in the optimization of the **life cycle**. Putative ancestors of land plants, charophyte green algae, did not have any sporophyte as their zygote proceeds to meiosis almost immediately after fertilization. New sporophyte could arise in connection with the need to disperse the spores from plants growing in the shallow water, where the wind acted as the most efficient dispersal agent. First, sporophytes served likely only for the storage of the haploid spores, but later most of the gametophyte functions were transferred to the sporophyte.

It is important to note also that the colonization of the land by plants was to happen after the formation of **soil**, the process involved bacteria, fungi, and invertebrates. Furthermore, the term "colonization of land" is not accurate since the actual land in the usual sense in those days did not exist; it was, in fact, huge, often completely flooded this wetlands-sea bottom space, interspersed with rock formations; there were no permanent freshwater. We can say that *animals and plants made the land themselves*, stopping erosion that once ruled the earth's surface. Land type familiar to us was formed slowly; we can, for example, assume that until Jurassic watersheds were completely devoid of vegetation.

1.5   First Life on Land

This epoch (spans Silurian and Devonian periods) began more than 440 million years ago and took about 85 million years. The Earth's climate was gradually warmer, starting with a small glaciation of Gondwana (the South Pole was in Brazil), climatic situation slowly reversed, and during the Devonian period, the world was dominated by abnormally high temperatures and extremely high ocean level. This time was ended with Caledonian orogeny, the result of proto-North America and proto-Europe collision, when mountains of Scandinavia, Scotland, and eastern North America have risen.

On land, there was a **radiation** (i.e., evolution in different directions) of terrestrial plants. There were already several **biomes**: bog communities, semi-aquatic ecosystems, and more dry plant associations with domination of mosses. Once the plants have "learned" how to make chemicals that make their cell walls much stronger (**lignin** and **suberin**), they started to make "skyscrapers" to escape competition for the light; this allowed them to grow up to the almost unlimited height. By the end of epoch, first **forests** appeared, which consisted of marattioid ferns (Fig. 2.11, middle left), giant horsetails, mosses, and first **seed plants**.

Origin of seed was most likely connected with the origin of trees. Ancestors of the seed plants (it is possible that they were close to modern tongue ferns, Fig. 2.11) were among the first plants to acquire the **cambium**, "stem-cell" tissue, and, consequently, the ability of the secondary thickening their trunk. After that, growth in height was virtually unrestricted. But there was another problem: the huge ecological gap between the giant sporophyte and minuscule, short-lived gametophyte dramatically reduced protection capabilities of the sporophyte and the overall plants' viability (a similar thing happened with dinosaurs in the Late Cretaceous).

Seed plants solved the problem and found the room for gametophyte right **on the sporophyte**. However, this change required plenty of coordination in the development (e.g., pollination), and initially, seed plants (like contemporary ginkgo, Fig. 2.12, top left, and cycads) were not much better than their sporic competitors.

At the seas, predatory vertebrates, armored fish "pushed" the old dominants, chelicerates (Fig. 2.26, bottom right) into the land. The last group became **first terrestrial predators**. There was already plenty of prey in the terrestrial fauna, in particular, millipedes and wingless proto-insects (Fig. 2.29, the middle). The last group (in order to escape predators) was likely forced to migrate to live **on trees**, and true insects appeared in the next epoch.

Shallow-water communities were dominated by advanced fish groups. The most important were ancestors of terrestrial vertebrates, lobe-finned fish (Fig. 2.17, 4th from top). These predatory animals, probably in order to "catch up" with the retreating water (as the tides at that time apparently extended for kilometers into the "land"), and also in the search for more food, started to develop adaptations to the terrestrial lifestyle. At the end of the epoch, they "made" organisms similar to modern amphibians, labyrinthodonts. They had many characters of terrestrial animals but likely spent most of their life in water.

From that time, most of the changes in marine ecosystems were only regrouping, reduction or increase of particular group. As an example, corals (existed from Cambrian) started to become the main builders of bioherms (as they are now). The role of trilobites decreased. Among mollusks prevailed ammonoids and nautiloids, cephalopods (Fig. 2.21, bottom right) with heavy shells. Vertebrates were represented not only with lobe-finned but with many other groups of fish-like animals, including jawless and different cartilaginous and boned fish. This epoch is often called "**the age of the fish**."

1.6 Coal and Mud Forests

Nikolaj Zinovjev, "Carboniferous Period" (1968)

This epoch took about 60 millions of years and is often described as the kind of tropical world, with warm and humid climate, plenty of CO_2 in the atmosphere, and the predominance of ferns. In fact, in the world at that time, the climate was quite variable. For example, the Arctic continent Angarida (or "Siberia", it corresponds with recent East Siberia) had really cold and dry climate. In contrast, the Euro-North-America was on the equator and had a tropical climate.

However, there was a little carbon dioxide and lots of oxygen; in fact, much more oxygen then it was on the whole history of Earth, both earlier and later. One of the proofs is an existence of giant **palaeodictyopteroidean insects**, some of them had more than a meter wingspan! As insects depend on the tracheal system for ventilation, it is safe to guess that there were plant of oxygen in the atmosphere to supply these big bodies.

The raise of oxygen is probably explained with appearance of forest biomes. Accumulation of coal is also related, the more carbon accumulated, the less should go into CO_2.

These Carboniferous forests were dominated with primitive woody ferns, tree-like horsetails, and basal seed plants (they have quite misleading name "seed ferns" but in fact, belonged to groups which now include ginkgo and cycads). There were also related to conifers (**cordaites** and, finally, **woody lycophytes** which now exist only as small water quillworts (*Isoëtes*).

Forests of this epoch were peculiar, and more similar to mangroves then to "normal" forests. They were systematically flooded with the tides and surf waves, and at the same time, decomposition of organic matter was slow (as there were no phytophagous insects and little fungi). Consequently, the bottom of such forest was probably covered with mud. This mud was threaded with numerous rhizomes of woody lycophytes.

They, as many other trees of these times, had imperfect thickening, and sooner or later would break and fall. Besides, sporic trees had no control over their microscopic gametophytes, and this resulted in periodical outbursts when many young plants of the same age started to compete and eliminate each other. All of these factors add to the existing mess, and lower levels of these forests were literally inundated with large size wood litter.

This was the primary ground of the origin of reptiles and flying insects. These two groups could origin "together", as elements of the one food chain also included trees. At the beginning of their evolution, many insect groups probably feed on generative organs of plants. Then dragonflies formed the first flying predators, and as the response, cockroaches and crickets went into the litter layer.

Some amphibians slowly evolved toward feeding on terrestrial invertebrates (like insects, slugs, and millipedes), and as a consequence, developed the full independence from water. This independence required substantial restructuring of the organization, in particular, the improvement of the respiratory system, skin, fertilization, and embryogenesis. Together, these changes resulted in the appearance of a new group, *reptiles*.

Seas in this epoch were dominated by mollusks, primitive arthropods, cartilaginous, and lobe-finned fishes.

1.7 Pangea and Great Extinction

Nikolaj Zinovjev, "Permian Period" (1969)

At the end of the Carboniferous period, there have been several important events. Firstly, all Earth continents collided in a single continent Pangea. Second, an active mountain-building started; this orogeny formed Urals, Altai, the Caucasus, Atlas, Ardennes. Then part of Pangaea (namely, Australia) "drove" to the South Pole, thus started the Great glaciation. Temperatures on Earth were thus even lower than it is now, in the epoch of Great Cenozoic glaciation.

Interesting is that these processes were not strongly affected the evolution of the biosphere, at least in the beginning. Of course, there are were new types of vegetation, conifer forests, savannas, and deserts. Three ferns declined, cycads (rare now) appeared. However, the fauna has not changed. The role of reptiles increased significantly, many of them were insectivorous, and some reptiles (synapsids) started to acquire characters of the future mammals. Amphibian stegocephalians have still thrived. Higher insects (insects with metamorphosis) were close to modern Hymenoptera and lived on conifers, and they played an essential role in the further evolution of the seed. In a forest, litter lived multiple herbivorous and predatory cockroach-like insects.

Reptile metabolism is entirely compatible with water life, so in Permian, some reptilian groups "returned" to water (this process continued in Mesozoic): there were marine, fish-eating mesosaurs, and freshwater hippo-like pareiasaurs.

At the end of the Permian period, about 270 million years ago, glaciation stopped. However, orogeny intensified, half of Siberia were covered with volcanic lava (famous Siberian Traps). That event probably was the reason for the *great extinction of marine life*: trilobites did not survive Permian, as well as 40% of cephalopods, 50% echinoderms, 90% brachiopods, and bryozoans, almost all corals and so on. More or less happily escaped were only sponges and bivalvians. However, some groups appeared first at this time, for example, contemporary bony fishes and decapod crustaceans.

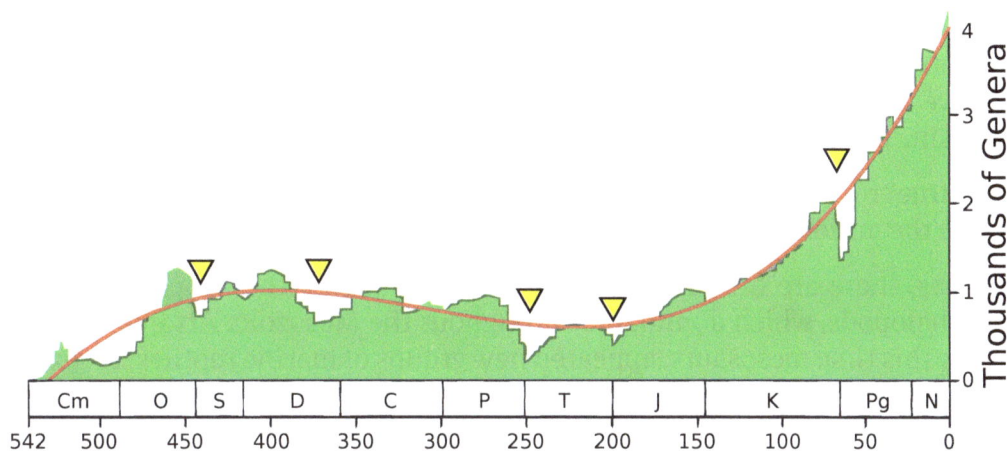

Five "great" extinctions. Great extinction of marine life is the third triangle from left.

1.8   Renovation of the Terrestrial Life

In the Triassic and early Jurassic, Pangea begins to disintegrate. The Atlantic Ocean (which still grows) opened. The climate was warm at first but very dry, and by the end of the era, it gradually became more convenient to the terrestrial life.

Among the seed plants, there appeared more advanced groups like bennettites, which participate in making savanna type vegetation (without grasses, though, the role of grasses was likely played with ferns, mosses, and lichens). Seeds of many plants were protected by scales or were embedded in an almost closed cupula. Seed protection was the "answer" of seed plants to the appearance of numerous phytophagous insect groups. Some other groups of insects began to adapt to the pollination of seed plants; this was an additional factor to facilitate the growing of seed covers.

Reptilians were still dominated but gradually replaced with various groups of archosauromorphs, the most advanced reptiles by that time, able to move very quickly, typically using only two legs.

Simultaneously run there were processes of "mammalization" and "avification" of reptiles. Ancestors of mammals were now in a small dimensional class and became insectivorous; this is because small herbivorous reptiles were simply physiologically impossible. Plant food is not very nutritional, and reptile feeding apparatus was unable to extract enough calories to support small, presumably more active animal. Giant herbivorous reptiles have less relative surface and therefore need fewer calories. Only turtles are an exception because of their "super-protection", which however has closed all further ways to improve the organization.

Ancestors of mammals were animals of the size of a hedgehog or less; they continued to improve their dental system, the thermal insulation system, and increase the size of the brain. The result was the emergence of first the first true mammals.

Among "true" reptiles, dinosaurs (birds' ancestors), crocodiles, and pterosaurs (which dominate the air for the next 70 million years) have appeared.

In the seas, there are first diatom algae, that stimulated the zooplankton, and in turn, cephalopods, which dominated throughout the Mesozoic. Also, to replace the extinct by this time mesosaurs, appeared new groups of marine reptiles, for example, notosaurs and molluscivorous placodonts.

1.9 Jurassic Park: World of Reptiles

The climate on Earth in this epoch (Jurassic and Early Cretaceous) approached the optimum, the split of the continental plates led to its humidification. A new flourishing of fauna and flora began. The sea strongly prevailed over the land, even high continental platforms such as the Russian and North African, were flooded.

The abundance of phytoplankton and zooplankton caused the thrive of marine fauna, including sponges, corals, bivalve mollusks (who took an active part in the construction of bioherms), echinoderms, etc. Ichthyosaurs and plesiosaurs were the most abundant marine predators.

Interestingly that in fossil deposits, pregnant females of ichthyosaurs are often found. Therefore, the ichthyosaurs were not only viviparous but also gave birth in conditions that "promoted" fossilization. The reason is they likely could not give birth as modern cetaceans: a tail up, this was not allowed with their vertical (like in fish, but not like in cetaceans) caudal fin. Then it seems that they were forced to give birth in shallow water, probably forming large groups (like modern seals).

On land, there were forests similar to the recent temperate taiga, composed mainly of representatives of the ginkgo class. Many of them were technically also angiosperms as their seeds were well protected by additional covers. These forests were mostly inhabited by insects, and primitive mammals hunted for them.

In open spaces, savanna forests were maintained (as modern grasslands exist only due to the constant pressure of ungulates) by giant herbivorous dinosaurs, replacing all the other groups with size of a modern cow and bigger. There also lived numerous predatory dinosaurs, both large and small bird-like insectivorous forms.

Flight of ancient birds was still very imperfect. The ancestors of birds needed feathers mainly for thermal insulation, and the flight occurred from the jumping movements required to catch flying insects. There is no much difference between archosauromorph reptiles and birds; in fact, flying is the only radical difference of birds.

The other group of flying archosaurs, pterosaurs, dominated the water and land borders. Ancestors of pterosaurs were fish-eating animals, and their flight arose as an adaptation to catching prey from the water.

1.10 The Rise of Contemporary Ecosystems

Nikolaj Zinovjev, "Cretaceous Period" (1969)

Cretaceous and Paleogene periods are usually referred to as different eras. However, here we join them in one epoch, as the development of the biosphere between the Cretaceous and the Paleogene did not change its direction.

The climate on Earth at that time was generally favorable for life, at in the end of the Cretaceous period, one of an absolute maximum of temperatures on Earth was observed. Continents gradually acquired current positions and outlines. **Alpine** orogeny began, then **Andes** and the **Rocky Mountains** arose, and then the **Himalayas**.

The main event of this epoch was the *Aptian revolution*. At the very end of the Lower Cretaceous almost simultaneously appeared those groups of animals and plants which are dominant to this day: flowering plants, polypod ferns, placental mammals, higher (tailless) birds, social insects (bees, ants and termites), butterflies, and higher bony fishes.

The origin of flowering plants for a long time was considered enigmatic. However, they do not radically differ from the rest of the seed plants: neither double fertilization nor protection of ovaries, much less the presence of a flower are unique attributes of flowering plants.

On the other hand, recent studies of both fossil and modern flowering plants indicate that the first flowering plants were *herbaceous perennials*, and some of them even aquatic. It is possible that during the previous epochs, some smaller primitive "gymnosperms", so-called "seed ferns" gradually acquired a herbaceous appearance, together with the capacity for easy vegetative reproduction ("partiality"), and a much shorter and more optimized life cycle. In the same direction, many other groups of seed plants were evolved, pushing each other's evolution, but the ancestors of flowering plants were the first to achieve this level.

Flowering plants colonized the land quickly, first at herbaceous stories where ferns and mosses could not compete with them (and there were no other seed plants, too). Then secondary woody flowering trees were formed, and apparently, they began to interfere with the woody "gymnosperms". By the end of the era, angiosperms forced out all other plants (except conifers) on the periphery of the ecosystems. As the climate gradually differentiated (becoming colder in high latitudes and warmer in the lower latitudes), tropical forests arose (they did not exist from the Carboniferous period).

An important event in the middle of the Upper Cretaceous was the occurrence of graminoids (grass-like plants). Capable of firmly retaining captured territory, they began to play an increasing role in communities.

The leaf litter of flowering plants, which is much copious than that of other seed plants (remember their fast life cycle), dramatically changed the carbon regime of freshwater ecosystems. Most of the oligotrophic (as modern sphagnum bogs) places have become mesotrophic or eutrophic, rich in organic substances. This is associated with strong changes in the fauna of insects (the emergence of higher forms of Diptera and beetles), and in turn, associated with the previous event the emergence of numerous insectivorous lizards, as well as with the radiation of tailed amphibians. Another consequence was probably a change of the outflow of some elements to the sea, possibly having an influence on the further development of the marine communities.

In the seas, various crocodiles, hampsosaurs, and giant mosasaur lizards dominated, and then extinct, likely due to the rapid radiation of fast-swimming higher bony fishes. At the end of the era, cetaceans appeared. Cephalopods began to decline, but the role of gastropods and bivalves significantly increased.

***

Extinction of dinosaurs is usually called the main event of this era. It must be said, however, that many dinosaur groups died out much earlier than the end of the Cretaceous, and many faded gradually, so Cretaceous extinction was only the "last stroke" of their decline.

On the other hand, the often-named *exogenous* causes of extinction (meteorite, etc.) do not explain why it touched practically only dinosaurs and had little effect on the evolution of the other tetrapods, insects and plants. In most of Earth history, exogenous influences cannot be firmly tied to any evolutionary event, for example, the time of the appearance of the largest meteorite craters of the Phanerozoic cannot be associated with any extinction.

One of the *endogenous* causes of extinction was the appearance of a predator capable of feeding on small and medium-sized prey (Rautian's hypothesis). The fact is that before the Cretaceous period, the animals of the small-sized class were represented only by insectivorous forms. However, gradually improving the dental apparatus, some mammals finally switched to plant food. These improvements finally led to the emergence of predatory forms capable of feeding on these herbivorous mammals. (Note that insectivorous animals of small size could not serve as regular food for any predator according to the law of the ecological pyramid.)

Since such a predator (they could be small predatory lizards, snakes, birds, and other mammals that appeared in this era) *could not be specialized only in one kind of prey*, it was necessarily the main enemy of small offspring of giant dinosaurs.

The other point is that the average size of adult dinosaurs increased dramatically by the end of the Cretaceous (this is the typical race of arms between prey and predator), but *young dinosaurs simply could not be large*! Dinosaur eggs had an upper limit of size because they (1) must be warmed to the center and also (2) be reasonably easy to hatch.

So small carnivores added much pressure to the gradual extinction of herbivorous, and after them, large predatory dinosaurs. Small dinosaurs evolve into birds, and whoever was left, did not have any significant advantages over mammals and birds, and therefore lost in the competition.

It is curious that the extinction of large predatory forms led to a kind of "vacuum" in terrestrial communities, and the most unexpected groups pretended to be predators before the advent of real predatory mammals (at the end of epoch): there were terrestrial crocodiles, giant predatory birds, and carnivorous ungulates.

Pterosaurs evolved into more and more large forms, and at the end of the era, they were unable to withstand competition with increasingly better flying birds. However,

22

the first flying mammals appeared: bats, whose flight arose, perhaps, as a means to save themselves from tree-ridden predators. Bats and birds safely divided the habitat, which is why they co-exist today.

Winning groups started extensive radiation. In the described epoch, several hundreds of order-level groups of mammals, birds and bony fishes appeared, and the most orders of flowering plants.

Nikolaj Zinovjev, "Tertiary" (1969)

1.11 Last Great Glaciation

Movements of continents in this epoch led to very adverse consequences. Panama and Suez isthmus closed, Antarctica gradually shifted to the area of the South Pole, and the northern continents surrounded the Arctic region as a ring. Everything now was ready for *new Great Glaciation*.

Life in the seas has not changed much. At the beginning of the epoch, due to the dryer climate and the progressive development of herbivorous mammals, *grasslands* were extensively expanded. These areas were inhabited by a fauna in which various proboscis, ungulates, rodents, and predatory mammals dominated.

One of the most curious episodes of this era was the *Great Inter-American Exchange*, the result of the formation of the Panama Isthmus. South America, isolated so far from all other continents (like Australia now), experienced the invasion of more advanced North American groups. Some South American animals have successfully withstood this onslaught and even advanced far to the north (opossums, armadillos, porcupines). However, the more significant part of the South American fauna went extinct.

After the formation of the glaciers, the rich Antarctic fauna and flora also died out, the last remnants (refugia) of which are now in the remnants of flooded Zealandia continent: now islands New Zealand, Lord Howe, and New Caledonia.

The advent of the glacier led to the formation of another type of community, *arctic steppe*: **tundra**, which advanced or retreated along with the ice.

The final accord of the development of the biosphere in this era was the appearance (most likely in East Africa) of representatives of the species *Homo sapiens* L.

Nikolaj Zinovjev, "Ice Age" (1969)

Chapter 2

Diversity of Life

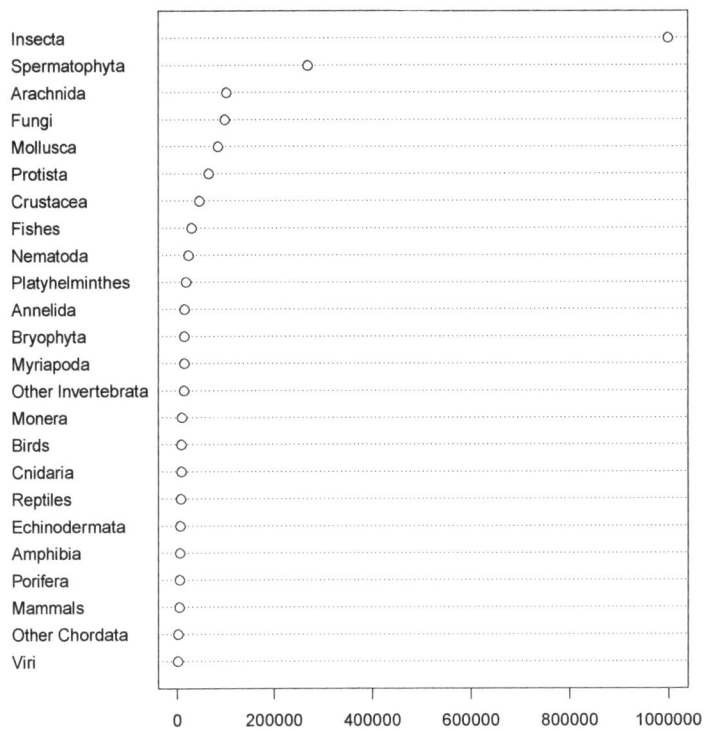

Figure 2.1: The most diverse groups of living things.

2.1 Diversity maps

animals

Ctenophora

Anthozoa: corals
Medusozoa: jelly fish

Placozoa

Porifera: sponges

Bilateria

Deuterostomia
Echinodermata:
starfish and others
Hemichordata
Chordata:
vertebrates

Acoelomorpha

Chaetognatha

Lophotrochozoa Nemertea
Mollusca:
Platyzoa: **gastropods**
flatworms and others
Lophophorata Annelida:
leeches and others

Ecdysozoa

Nemathelminthes
Arthropoda:
all arthropods

protists

Fungi and allies
Sulcozoa
Amoebozoa: **slime molds**
Choanozoa
Opisthosporidia
Eomycota
Entorrhizomycota
Basidiomycota: **mushrooms**
Ascomycota: **lichens**

Metamonada
Discoba

Algae and allies

Rhizaria

Labyrinthomorpha
Opalozoa
Oomycota
Chromophyta:
brown algae

Myzozoa
Ciliophora

Heliozoa
Haptophyta
Cryptista

Glaucophyta
Rhodophyta:
red algae
Chlorophyta:
green algae

prokaryotes

Archaea

Nanoarchaeota
Euryarchaeota
Filarchaeota
Asgardarchaeota

Bacteria

Firmicutes
Actinobacteres
Cyanobacteria
Chloroflexes

Elusimicrobia
Spirochaetae
Planctobacteria
Bacteroidetes
Proteobacteria

plants

Bryophyta:
**mosses
and allies**

Pteridophyta:
**ferns
and allies**

Spermatophyta:
**gymnosperms
and angiosperms**

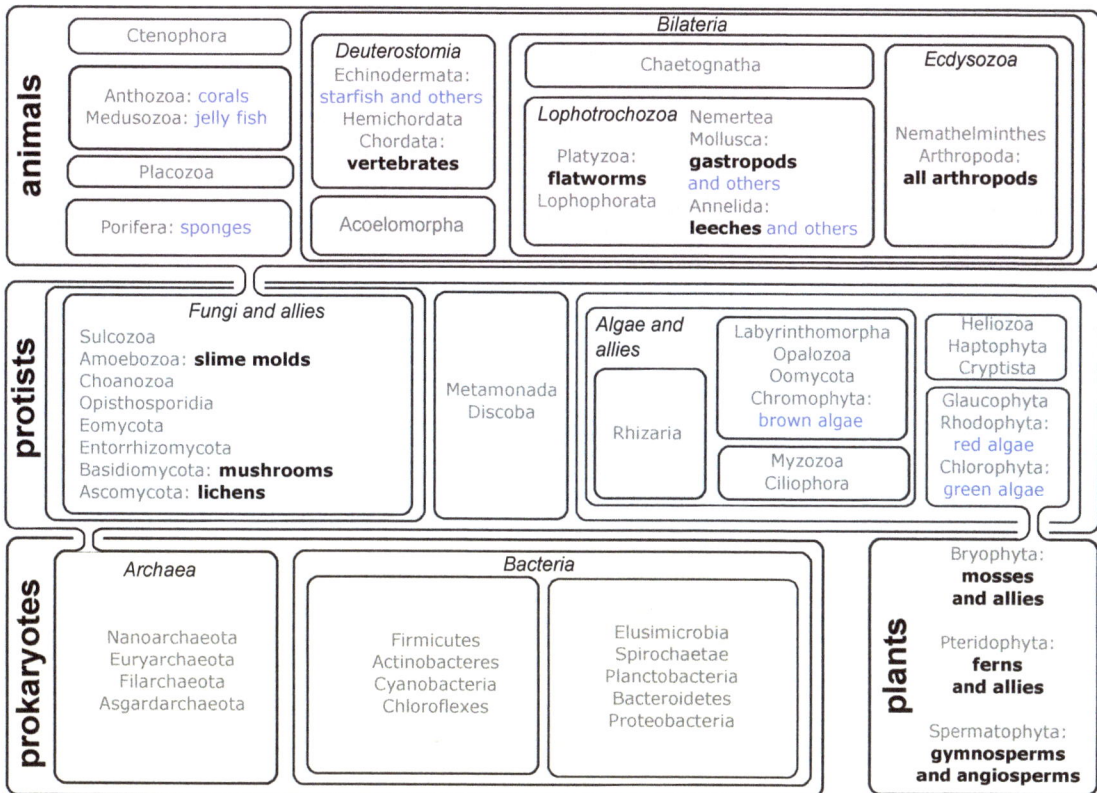

Treemap of Life.

Lobopoda	Tardigrada	**Arthropoda** s.l.:

Arthropoda s.str.:

Cheliceromorpha:

Chelicerata:

Arachnida:

Araneae: **spiders**
Acari: **ticks and mites**
Scorpiones
Amblypigi: whip spiders
Opiliones: **harvestmen**
... and smaller orders

Myriapoda:

Chilopoda: **centipedes**
Diplopoda: **millipedes**
...

Other crustacean groups

Mala-costraca:

Decapoda:
**crabs,
crayfish,
shrimp**
Isopoda:
woodlice

Pancrustacea:

Hexapoda: | Entognatha

Polyneoptera:

Orthoptera:
grasshopers
Mantoidea:
mantids
Blattoidea:
cockroaches
Isoptera:
termites
...

Paraneoptera:
Hemiptera: **bugs and aphids**
...

Insecta
Neoptera

Palaeoptera:
Odonata: **dragonflies**

Holometabola:

Hymenoptera:
**wasps
and ants**

Neuropteroidea:
Coleoptera:
beetles
Neuroptera:
lacewings
Megaloptera:
dobsonflies
...

Mecopteroidea:

Trichoptera:
caddisflies
Lepidoptera:
moths
Diptera:
flies
...

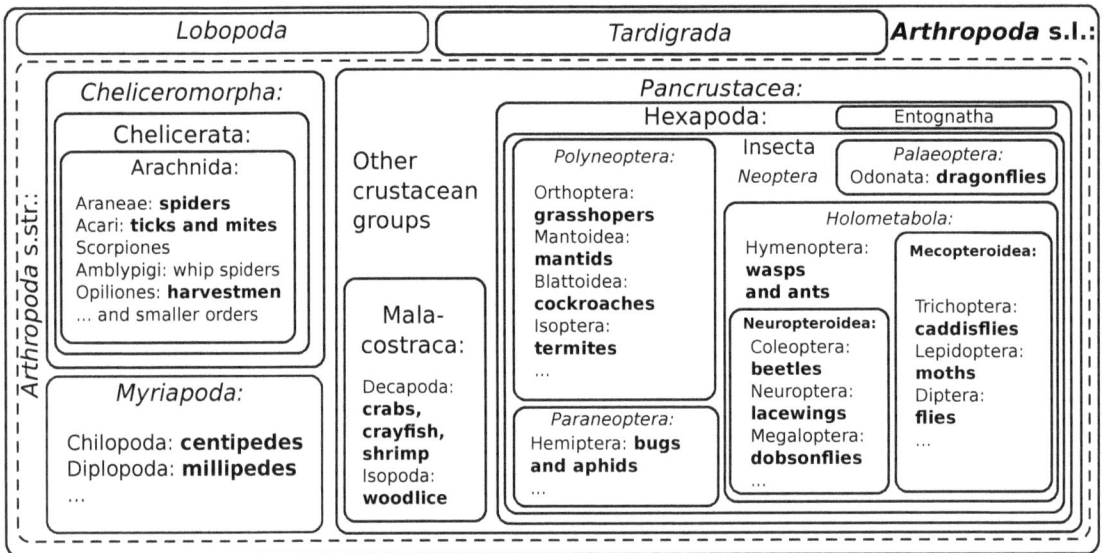

Treemap of arthropods.

Bryophyta:

- **liverworts** (Hepaticae)
- **mosses** (Bryophytina)
- hornworts (Anthocerotophytina)

Pteridophyta:

lycophytes: clubmosses, spikemosses, quillworts (Lycopodiopsida)

monilophytes (Pteridophytina):

- **horsetails** (Equisetopsida)
- **whisk ferns** (Psilotopsida)
- tongue ferns (Ophioglossopsida)
- giant ferns (Marattiopsida)
- **true ferns** (Pteridopsida)

seed plants (Spermatophyta):

"gymnosperms":

- **conifers** (Pinopsida)
- Gnetopsida
- **cycads** (Cycadopsida)
- ginkgo (Ginkgoopsida)

flowering plants (Angiospermae):

- **liliids**, monocots (Liliidae)

"dicots":

- **magnoliids** (Magnoliidae)
- **rosids** (Rosidae)
- **asterids** (Asteridae)

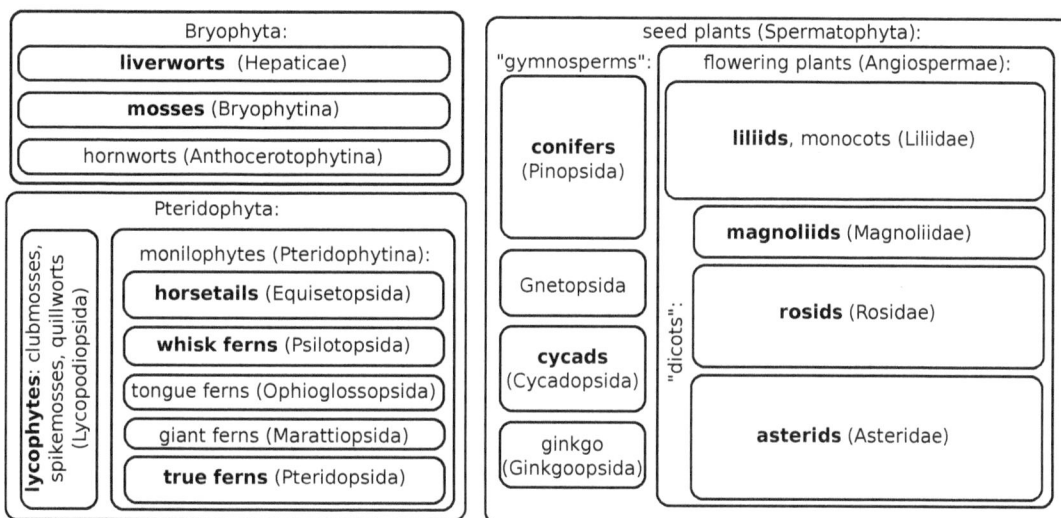

Treemap of plants.

asterids

Asterales

Compositae,
aster family

Lamiales
**Labiatae,
mask family**
Gesneriaceae

Solanaceae

Gentianales
**Rubiaceae,
coffee family**

**Apocynaceae,
dogbane
family**

Araliales

Ericales

Garryales

Cornales

Cucurbitales

Fagales

Begoniaceae

Zingiberales
Zingiberaceae
Musaceae

Caryophyllales

Dilleniales

Malpighiales

Fabales

Poales
**Gramineae,
grasses**

**Cyperaceae,
sedge family**

Berberidopsidales

Vitales

**Euphorbiaceae,
spurge family**

Leguminosae

Santalales

Celastrales

Arecales
Palmae

Alismatales

Zygophyllales

Rosales

Liliales

Araceae

Gunnerales

Oxalidales

**Urticaceae,
...**

Saxifragales

rosids

Huerteales

Ochidaceae

Piperales
Piperaceae

Geraniales
Picramniales

Magnoliales

magnollids

Ranunculales

Platanales

Sapindales

Brassicales

Chloranthales

Myrtales
Melastomataceae

Malvales
Malvaceae

Nymphaeales

Amborellales

"dicots"

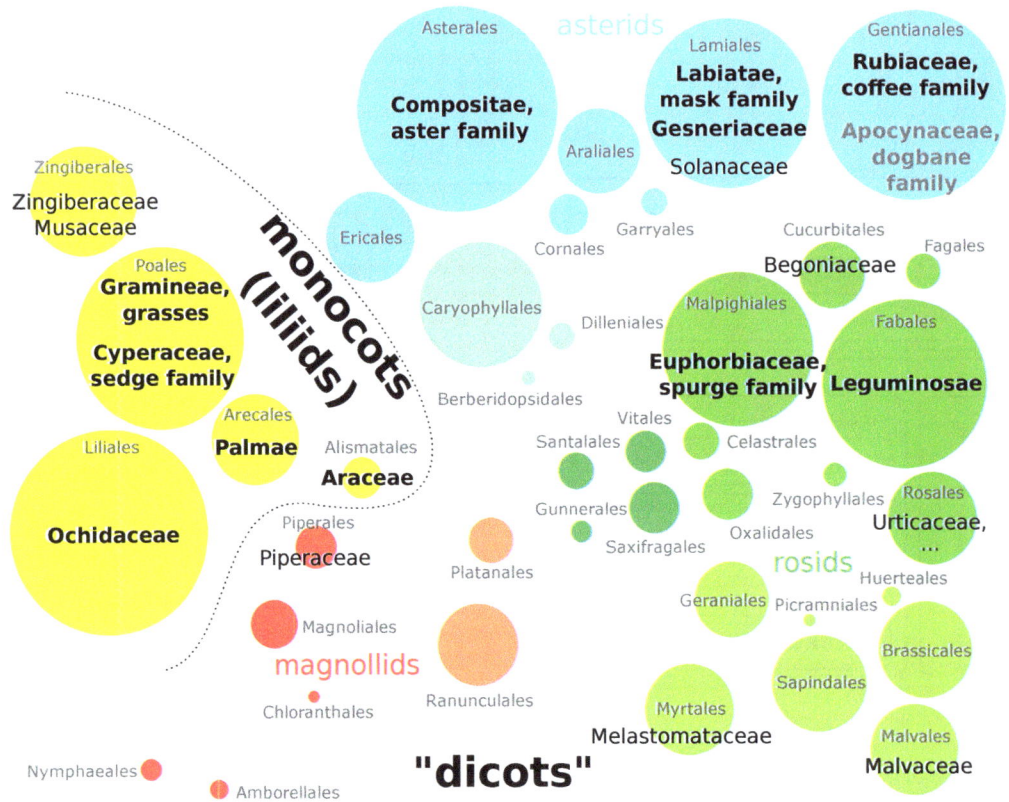

Treemap of angiosperms.

monocots (liliids)

"*Pisces*":

Agnatha (jawless fishes)

Chondrichthyes (cartilaginous): Elasmobranchii (**sharks** and **rays**) and Holocephala (**chimaeras**)

"*Osteichthyes*" (bony):

Dipnoi (*Sarcopterygii*, lobe-flnned): Ceratodontiformes, Lepidosireniformes, Coelacanthiformes (***latimeria***)

Actinopterygii (*Actinopterygii*, ray-flnned):

Chondrostei: Ascipenseriformes (**sturgeon, paddlefish**), Polypteriformes (bichir)

Neopterygii:

Holostei: Amiiformes (**bowfin**) Lepisostei-formes (**spotted gar**)

Teleostei:
Clupeiformes (**herring**)
Cypriniformes (**zebrafish**)
Characiformes (**tetras**)
Siluriformes (**catfish**)
Salmoniformes (**salmon**)
Gadiformes (**cod**)

Acanthopterygii:
Syngnathiformes (**seahorses**)
Tetraodontiformes (**pufferfish**)
Pleuronectiformes (**flatfish**)
Scorpaeniformes (**scorpionfish**)
Perciformes (**clownfish, tang, mackerel, cichlids, goby, ...**)

Treemap of fishes.

Aves:

Palaeognatha: **ostrich, cassovaries, tinamous**

Neognatha:

Galloanseres: **chicken and geese**

Neoaves

Opisthocomiformes: hoatzin

Mirandornites: **fiamingos, grebes**

Columbimorphae: **doves**

Otidimorphae: **cuckoos, turacos**

Caprimulgiformes: **hummingbirds, nightjars**

Telluraves:

Afroaves:

Accipitriformes: **vultures, eagles**

Cavitaves:

owls, trogons, hornbills, toucans, woodpeckers, kingfishers

Australaves:

seriemas, falcons, parrots,

passerines:

NZ wrens
***Suboscines*: cotingas, ...**
***Songbirds*: corvides, passerides, ...**

Aequornites:
shorebirds, pelicans, herons, cormorants, penguins, loons ...

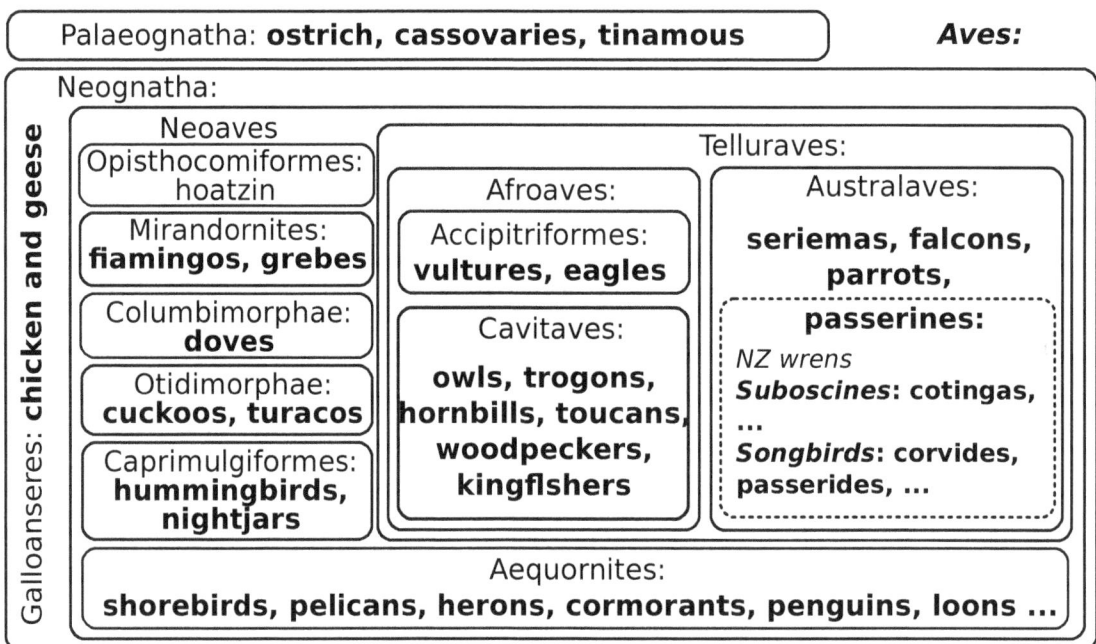

Treemap of birds.

Mammalia:

Prototheria: **platypus, echidna**

Theria:

Marsupialia: **pouched mammals**

Placentalia:

Atlantogenata:

Xenarthra: **anteaters, sloths, armadillos**

Afrotheria:

Macroscelidae

Afrosoricida

Tubulidentata: **aadvark**

Syrenia: **manatees**

Proboscidea: **elephants**

Hyracoidea: **hyrax**

Boreoeutheria:

Laurasitheria:

Eulypotyphla

Chiroptera: **bats**

Perissodactyla: **horses, rhinos, tapirs**

Cetartiodactyla: **pigs, whales, cows, camels**

Pholidota

Carnivora: **cats and dogs**

Euarchontoglires:

Scandentia: **tupaias**

Dermoptera

Primates: **us**

Lagomorpha: **rabbits**

Rodentia:
hystricomorphs (**aguti**), sciuromorphs (**squirrels**), castorimorphs (**beavers**), myomorphs (**mice**)

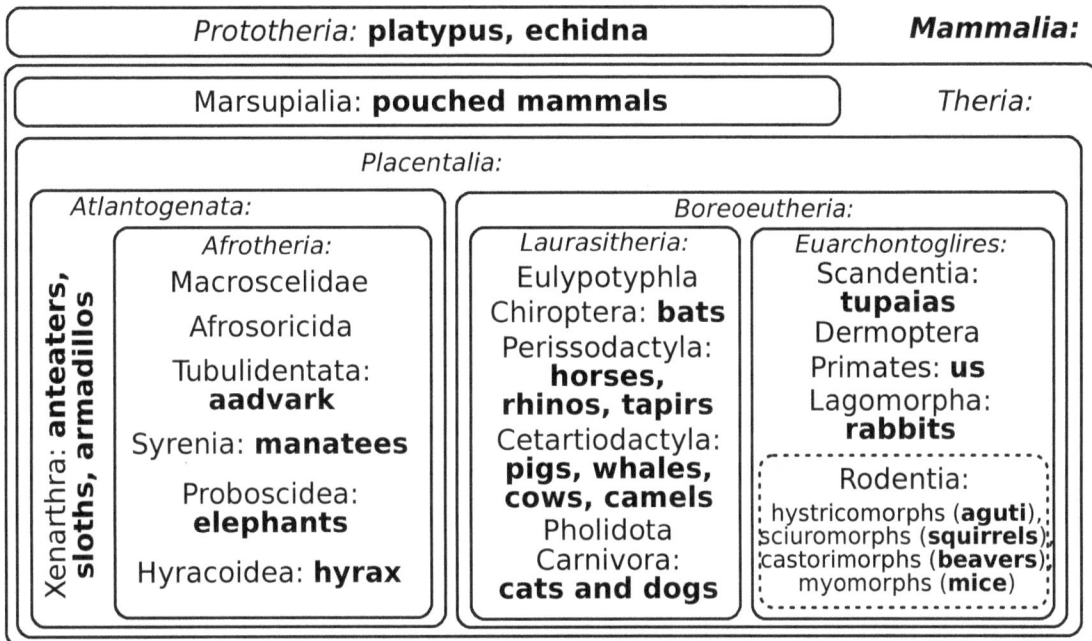

Treemap of mammals.

Treemaps placed above are alternatives to dendrograms ("phylogeny trees") and classification lists ("classifs"). Please remember that this is only one of ways to represent hierarchical, tree-like data. All three approaches are generally equivalent.

2.2 Illustrations

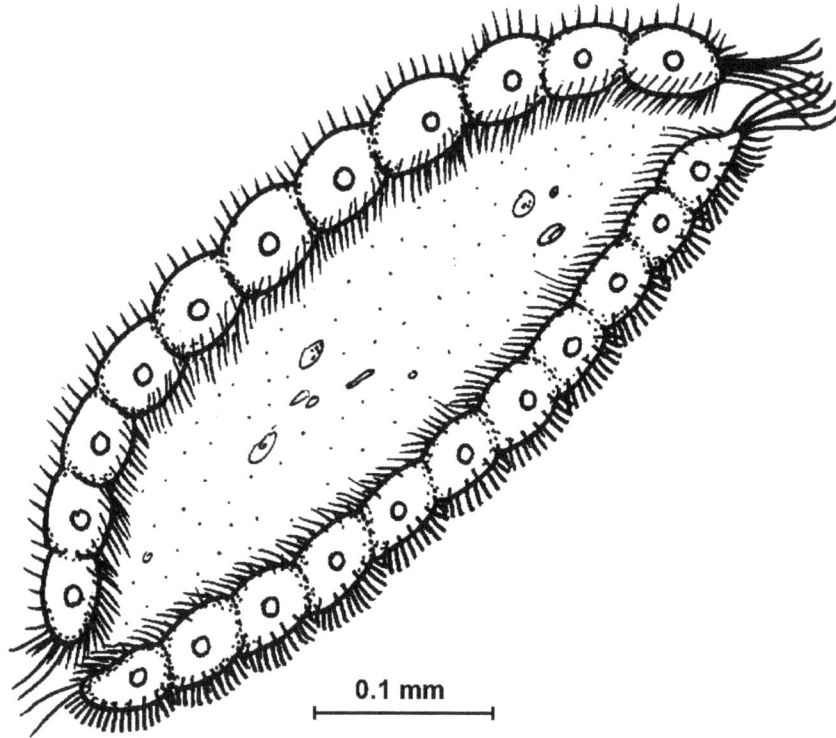

Salinella salve, the most mysterious living thing discovered ever
(slightly changed from the BIODIDAC original).

Below, all black and white illustrations were provided by Georgij Vinogradov and Michail Boldumanu.

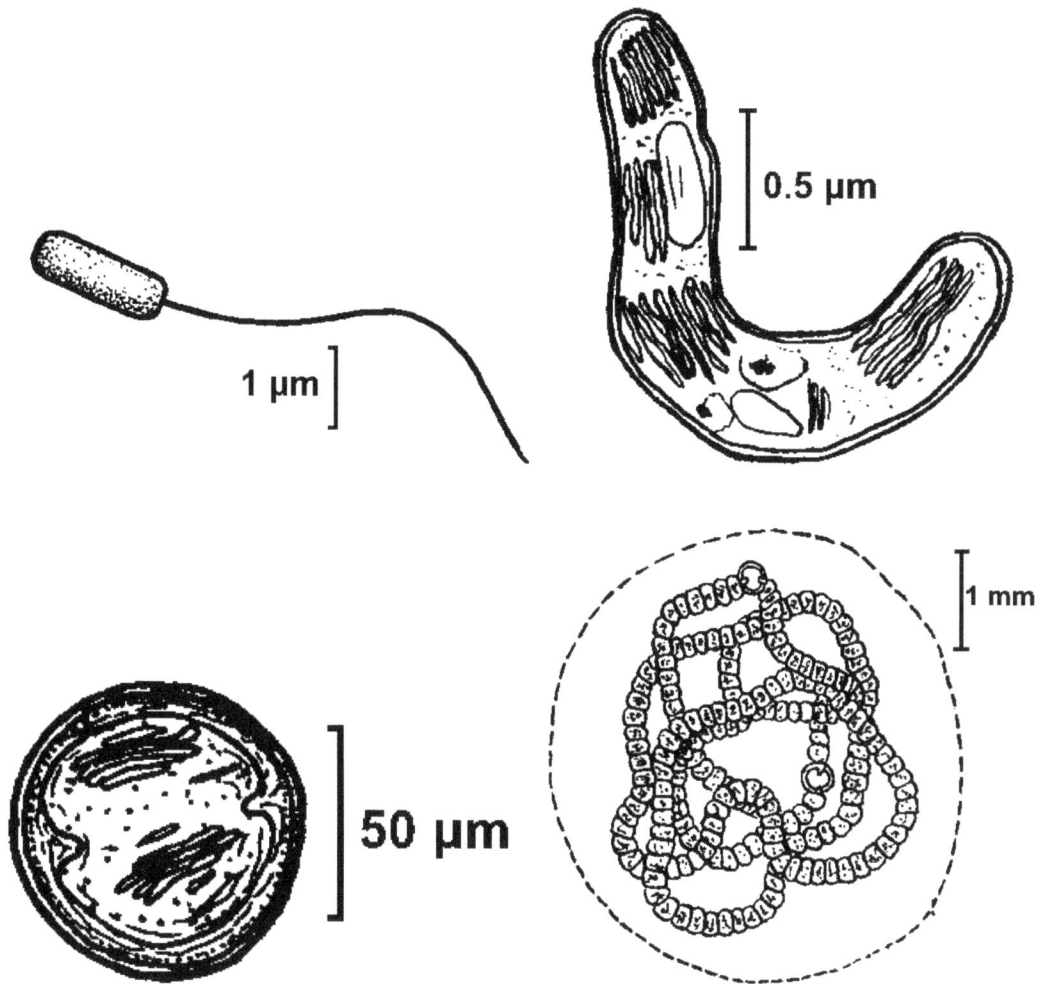

Figure 2.2: **Monera: Bacteria**. Left to right, top to bottom: Firmicutes: *Bacillus* sp.; Chlorobia: *Chlorobium* sp.; Cyanobacteria: *Prochloron* and *Nostoc kihlmanii*.

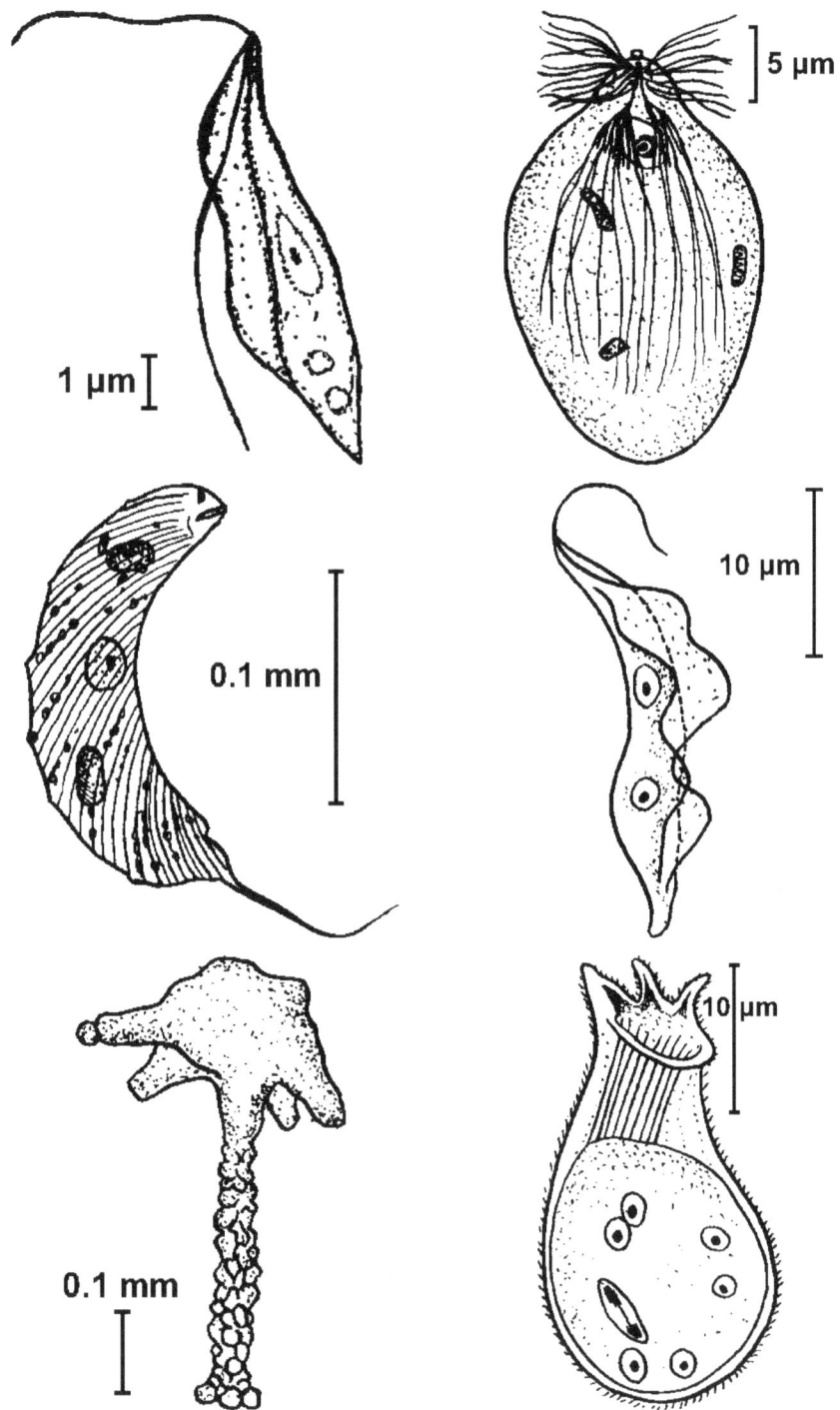

Figure 2.3: **Protista: Excavata.** Left to right, top to bottom: Jacobea: *Jakoba libera*; Parabasalea: *Barbulanympha ufalula*; Euglenophyceae: *Euglena spirogyra*; Kinetoplastea: *Trypanosoma brucei*; Heterolobosea: *Acrasis rosea* and *Stephanopogon colpoda*.

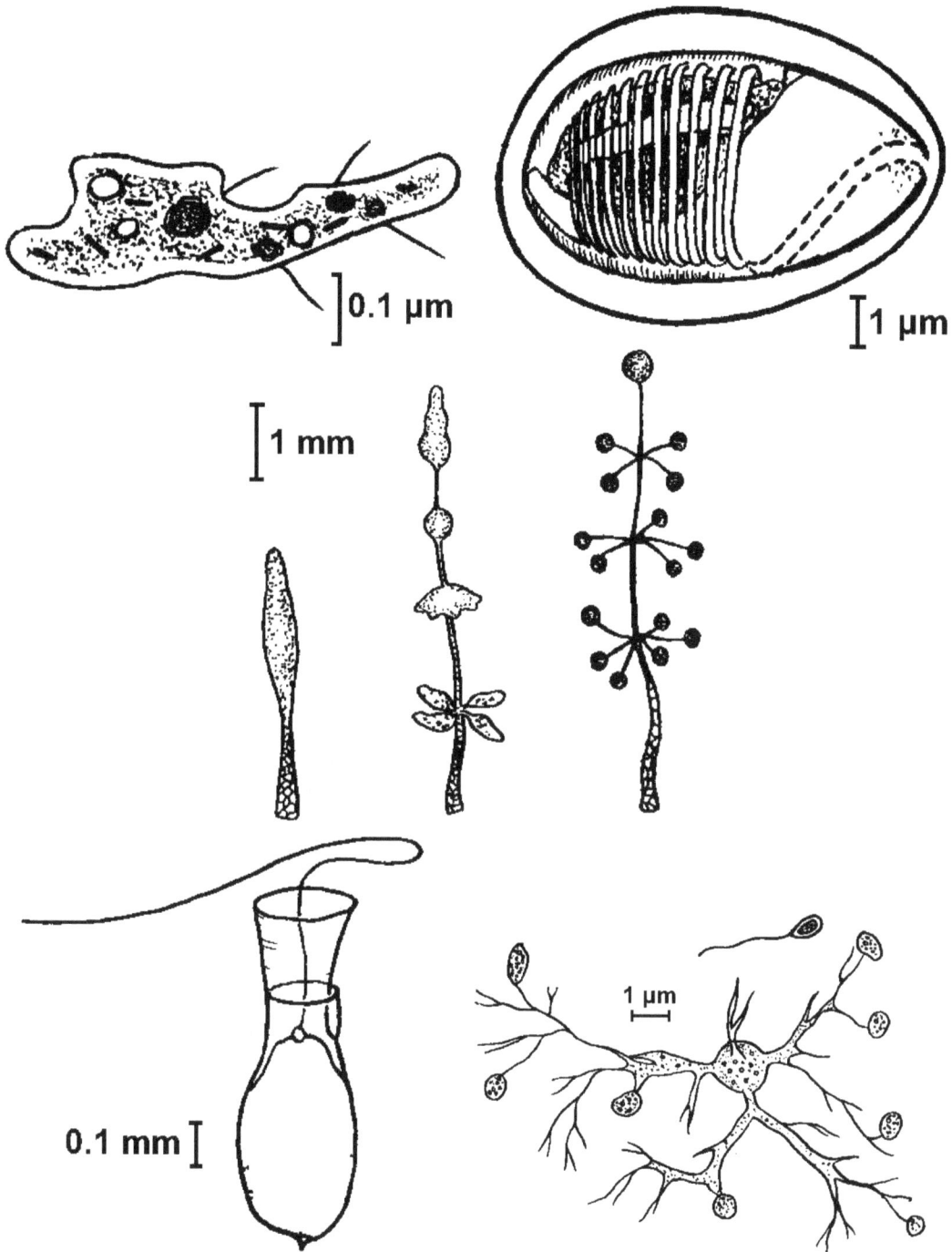

Figure 2.4: **Protista: Panmycota I**. Left to right, top to bottom: Archamoebae: *Pelomyxa palustris*; Microsporea: *Theloliania* sp. spore; Macromycerozoa: *Polysphondylium pallidum*; Choanomonadea: *Codosiga botrytis*; Chytridiomycetes: *Polyphagus euglenae*.

37

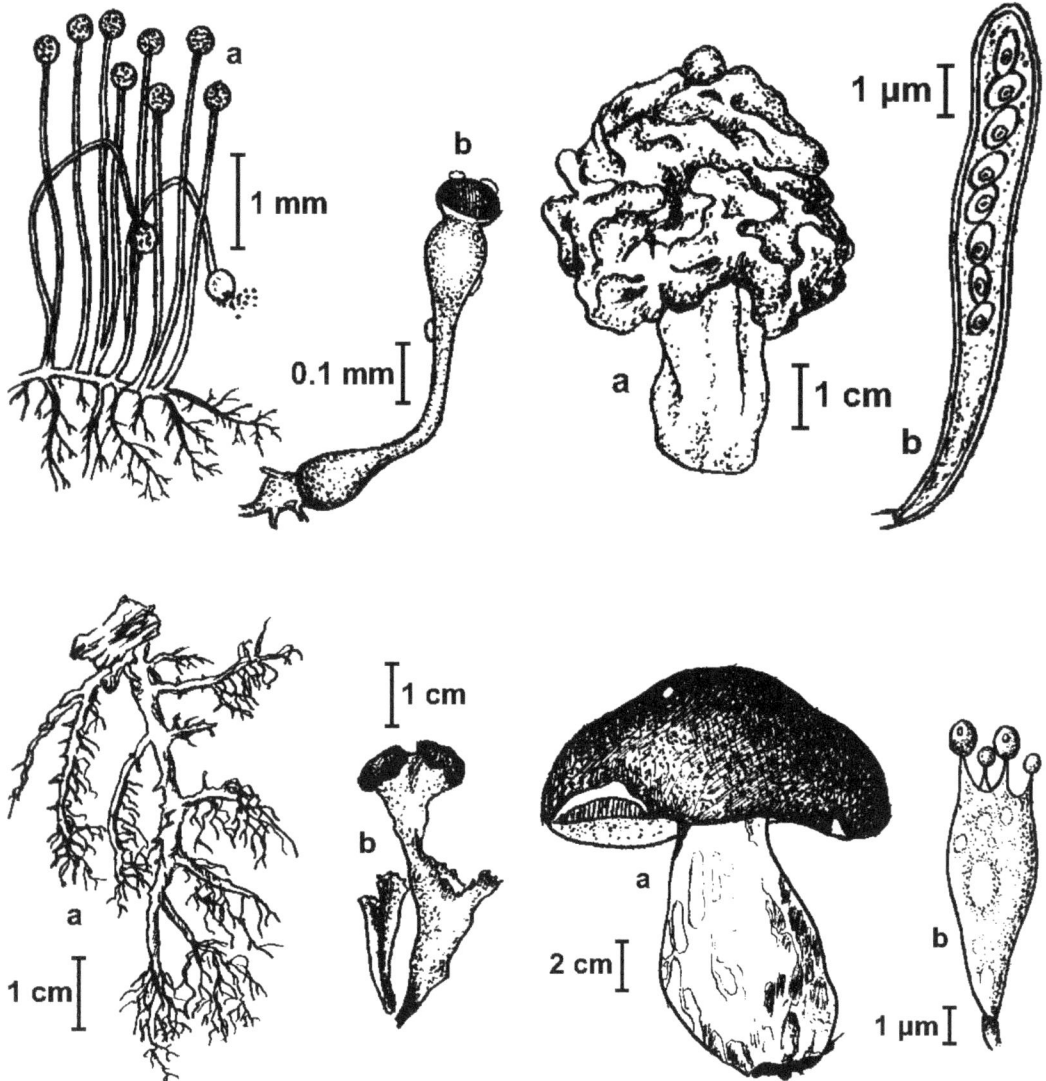

Figure 2.5: **Protista: Panmycota II**. Left to right, top to bottom: Zygomycetes: a *Mucor* sp., b *Pilobolus* sp.; Ascomycetes: *Gyromitra esculenta* (a fruiting body, b ascus) and lichenes: a *Usnea* sp., b *Cladonia deformis*; Basidiomycetes: *Boletus edulis* (a fruiting body, b basidia).

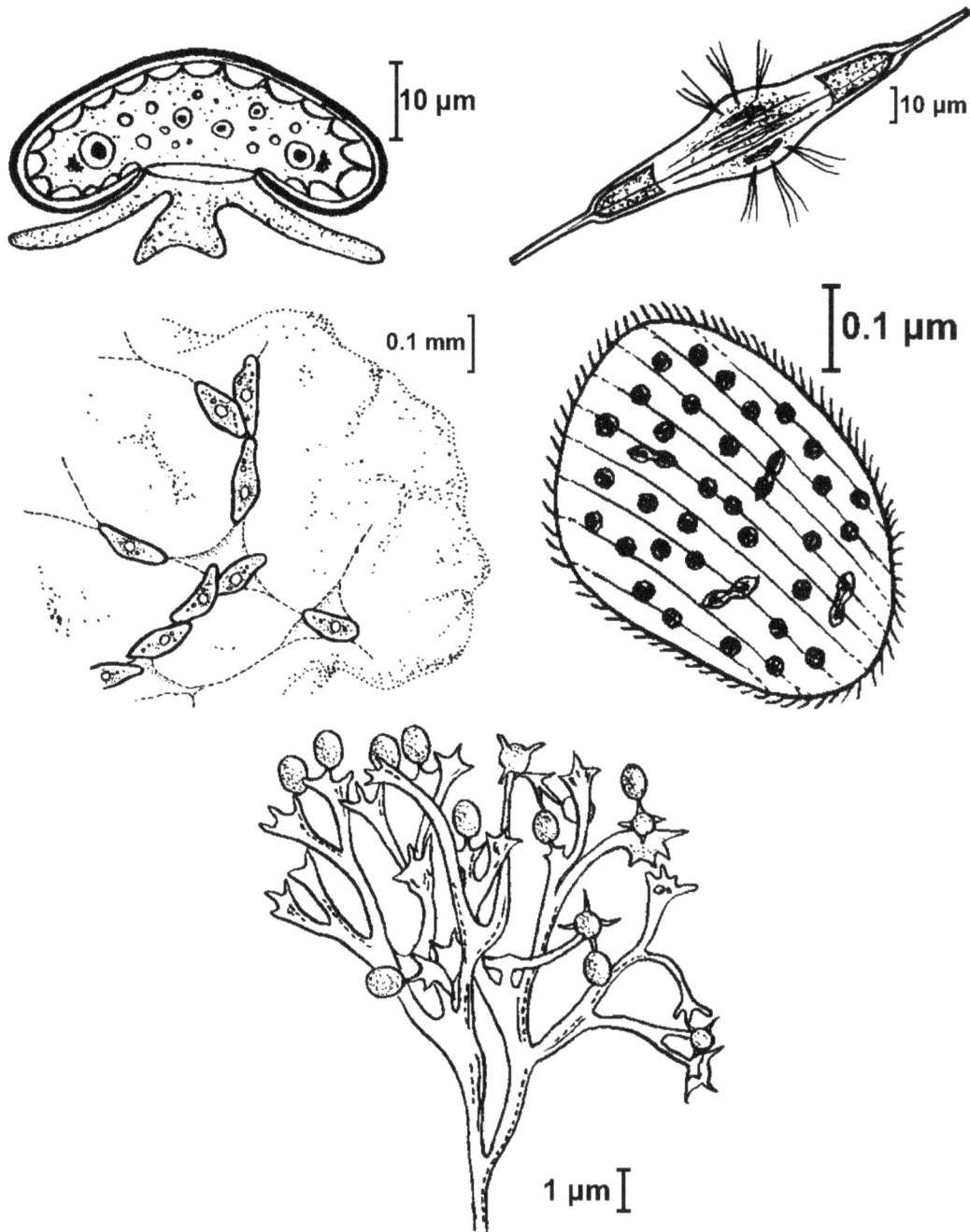

Figure 2.6: **Protista: Panalgae: Chromobionta I**. Left to right, top to bottom: Rhizopodea: *Arcella vulgaris*; Acantharia: *Amphilonche elongata*; Labyrinthulea: *Labyrinthula coenocystis*; Opalinea: *Opalina ranarum*; Oomycetes: *Bremia* sp.

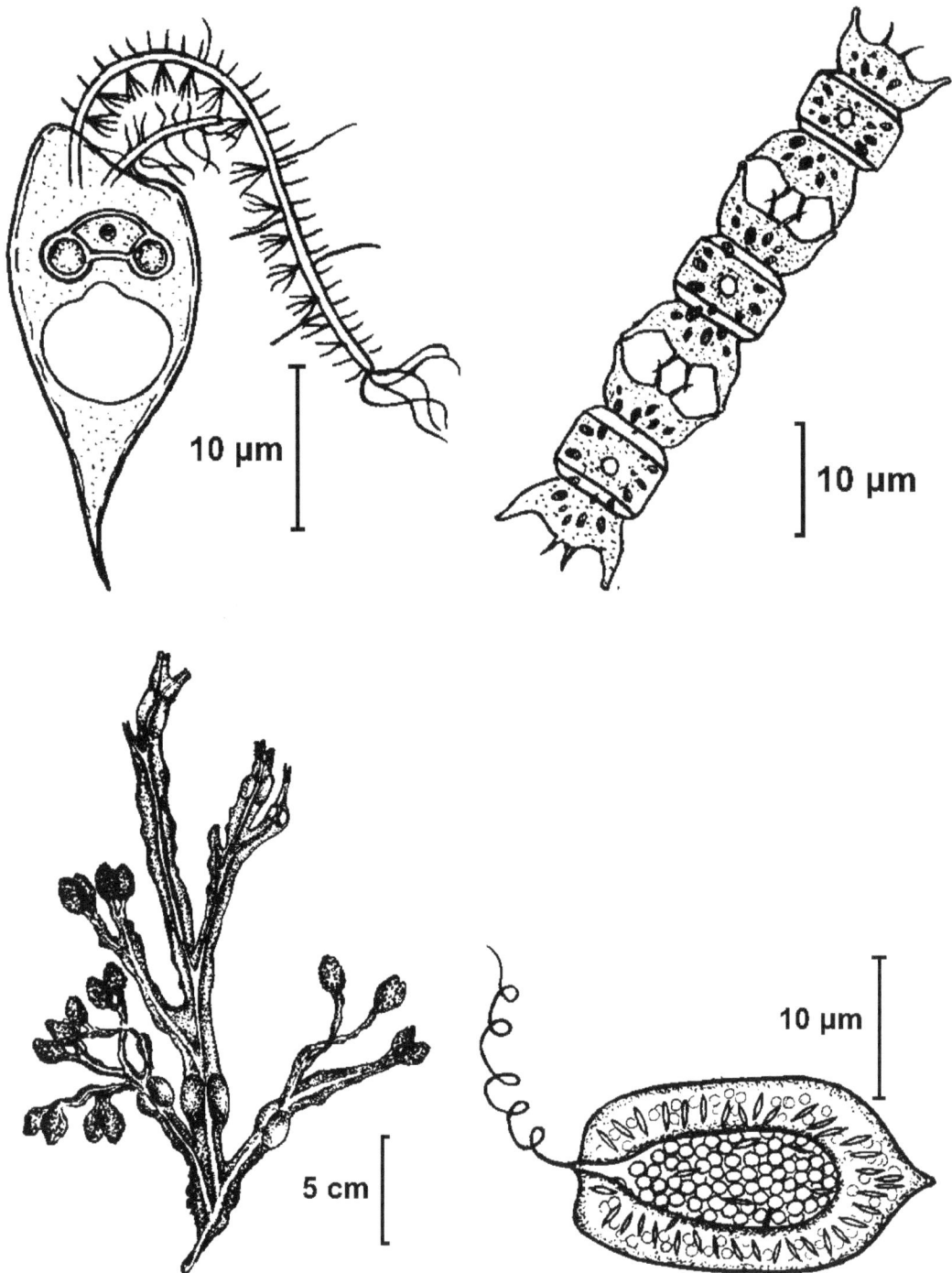

Figure 2.7: **Protista: Panalgae: Chromobionta II (Chromophyta)**. Left to right, top to bottom: Chrysophyceae: *Ochromonas danica*; Bacillariophyceae: *Biddulphia aurita*; Phaeophyceae: *Fucus vesiculosus*; Raphidophyceae: *Gonyostomum semen*.

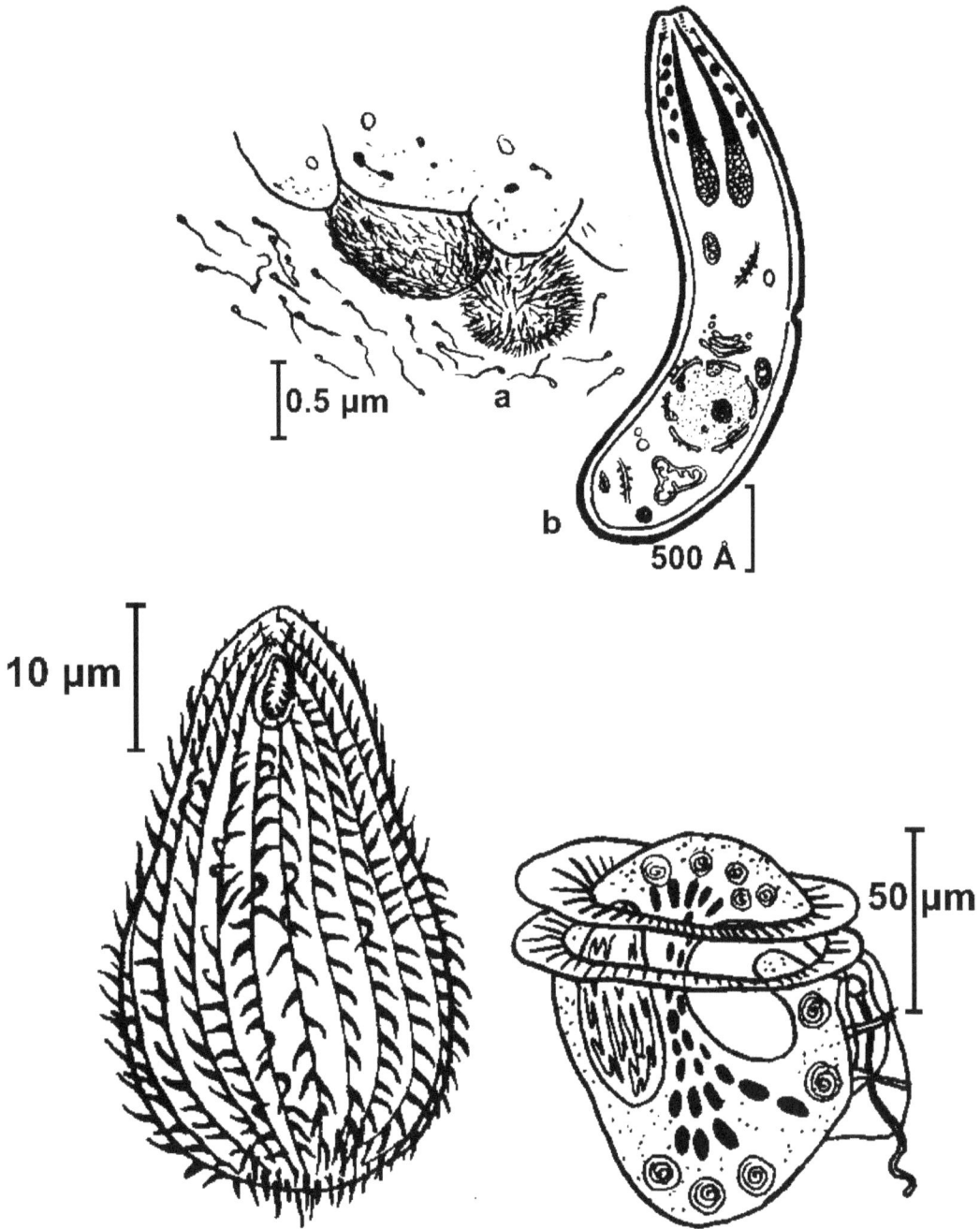

Figure 2.8: **Protista: Panalgae: Chromobionta III (Alveolata).** Left to right, top to bottom: Coccidiomorpha: *Plasmodium vivax* (a in host tissues, b separate sporozoite); Oligohymenophorea: *Tetrahymena pyriformis*; Dinozoa: *Phalacroma* sp.

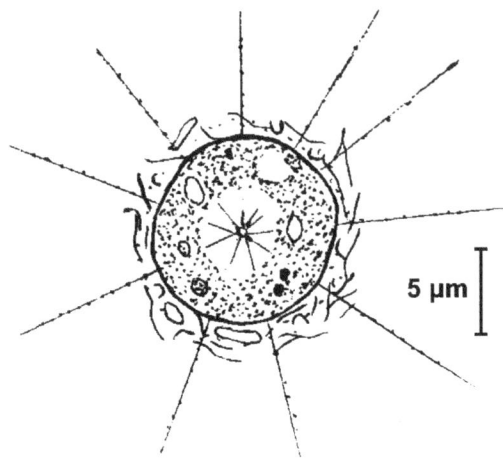

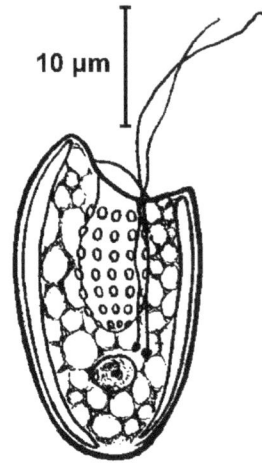

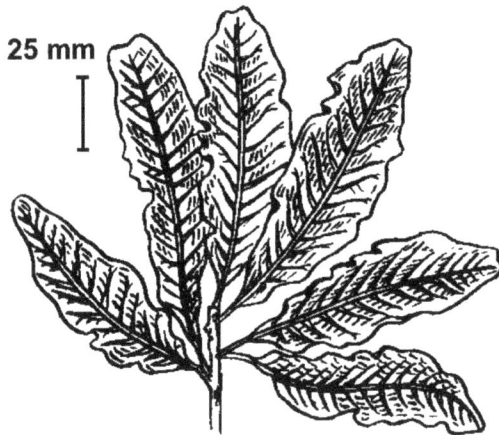

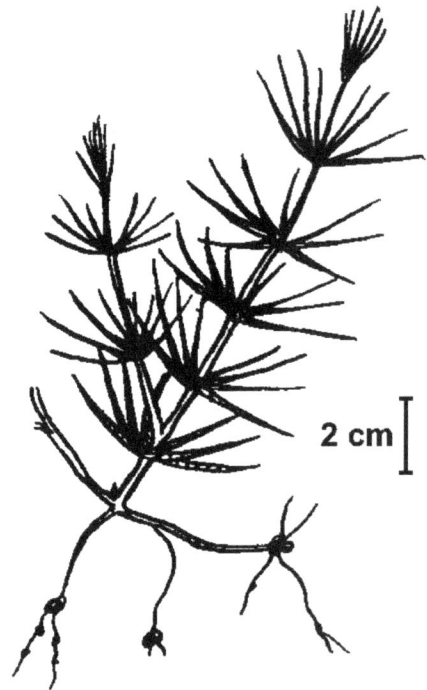

Figure 2.9: **Protista: Panalgae: Hacrobia and Chlorophyta**. Left to right, top to bottom: Centrohelea: *Raphidiophrys capitata*; Cryptophyceae: *Cryptomonas ovata*; Rhodophyta: *Delesseria* sp.; Charophyceae: *Chara fragilis*.

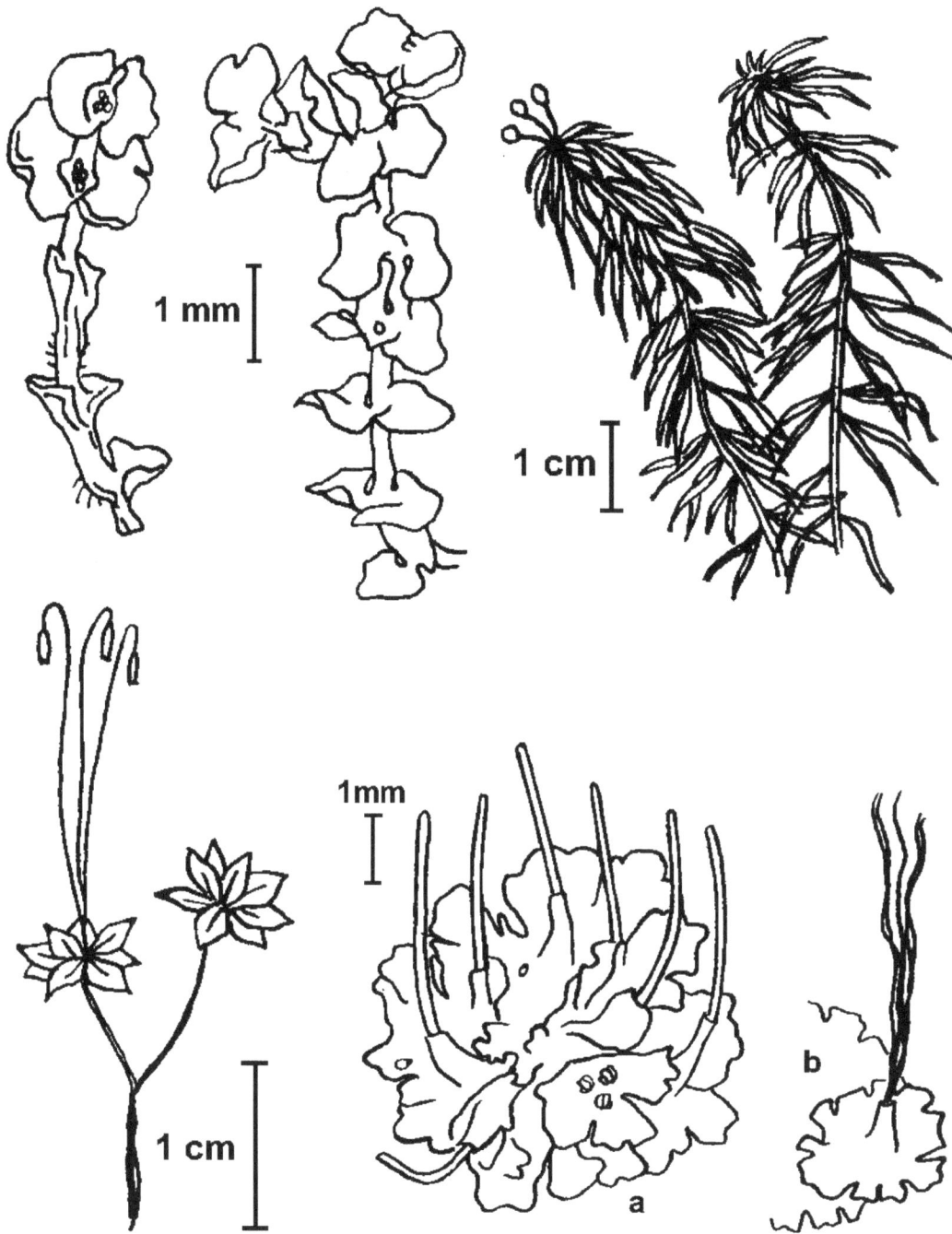

Figure 2.10: **Vegetabilia I (Bryophyta).** Left to right, top to bottom: Jungermanniopsida: *Phyllothallia* sp.; Sphagnopsida: *Sphagnum* sp.; Bryopsida: *Rhodobryum roseum*; Anthocerotopsida: *Anthoceros laevis* (a open sporogon, b general view).

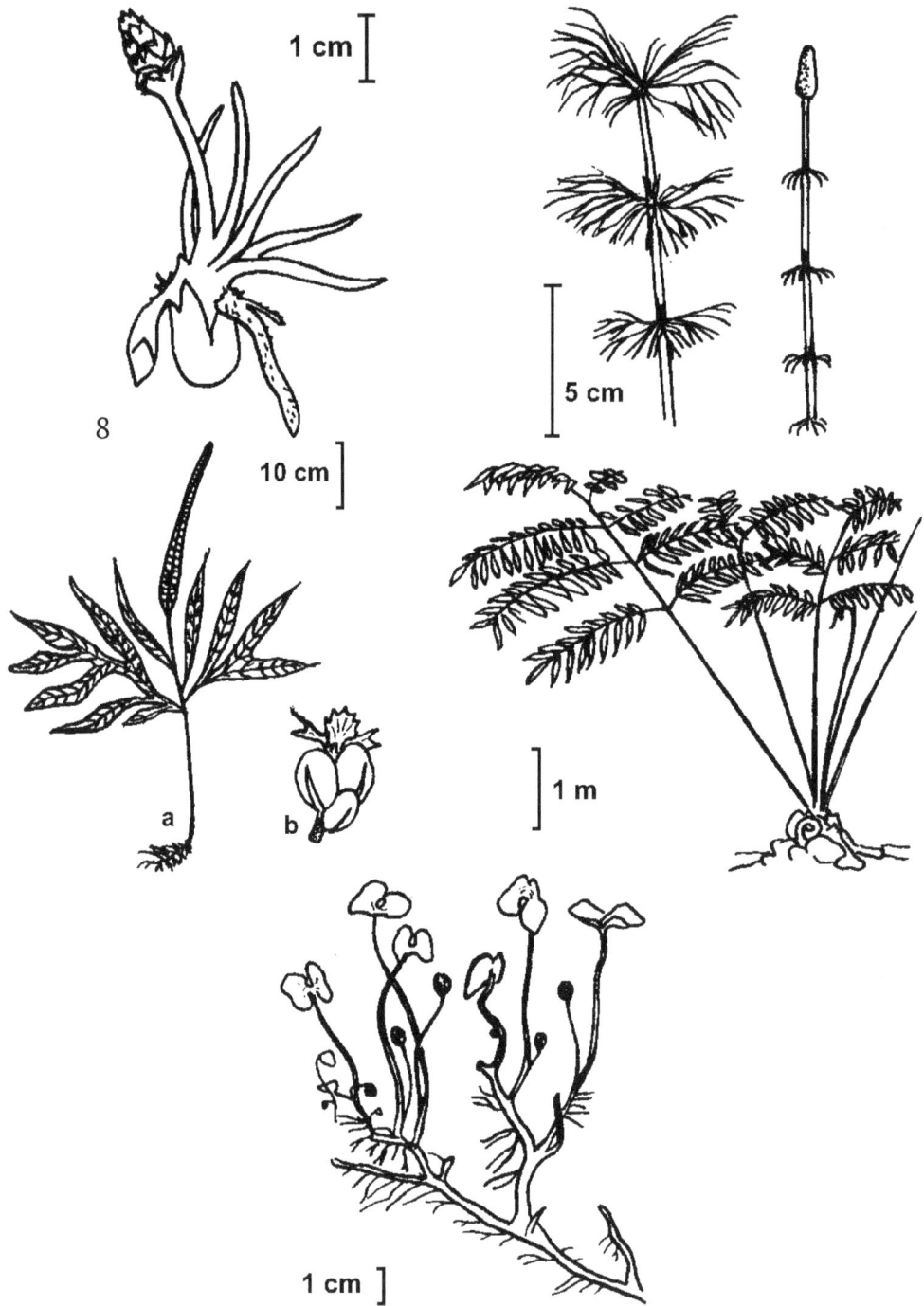

Figure 2.11: **Vegetabilia II (Pteridophyta)**. Left to right, top to bottom: Lycopodi-opsida: *Phylloglossum drummondii*; Equisetopsida: *Equisetum sylvaticum*; Ophioglos-sopsida: *Helminthostachys zeylanica*; Marattiopsida: *Angiopteris evecta*; Pteridop-sida: *Regnellidium diphyllum*.

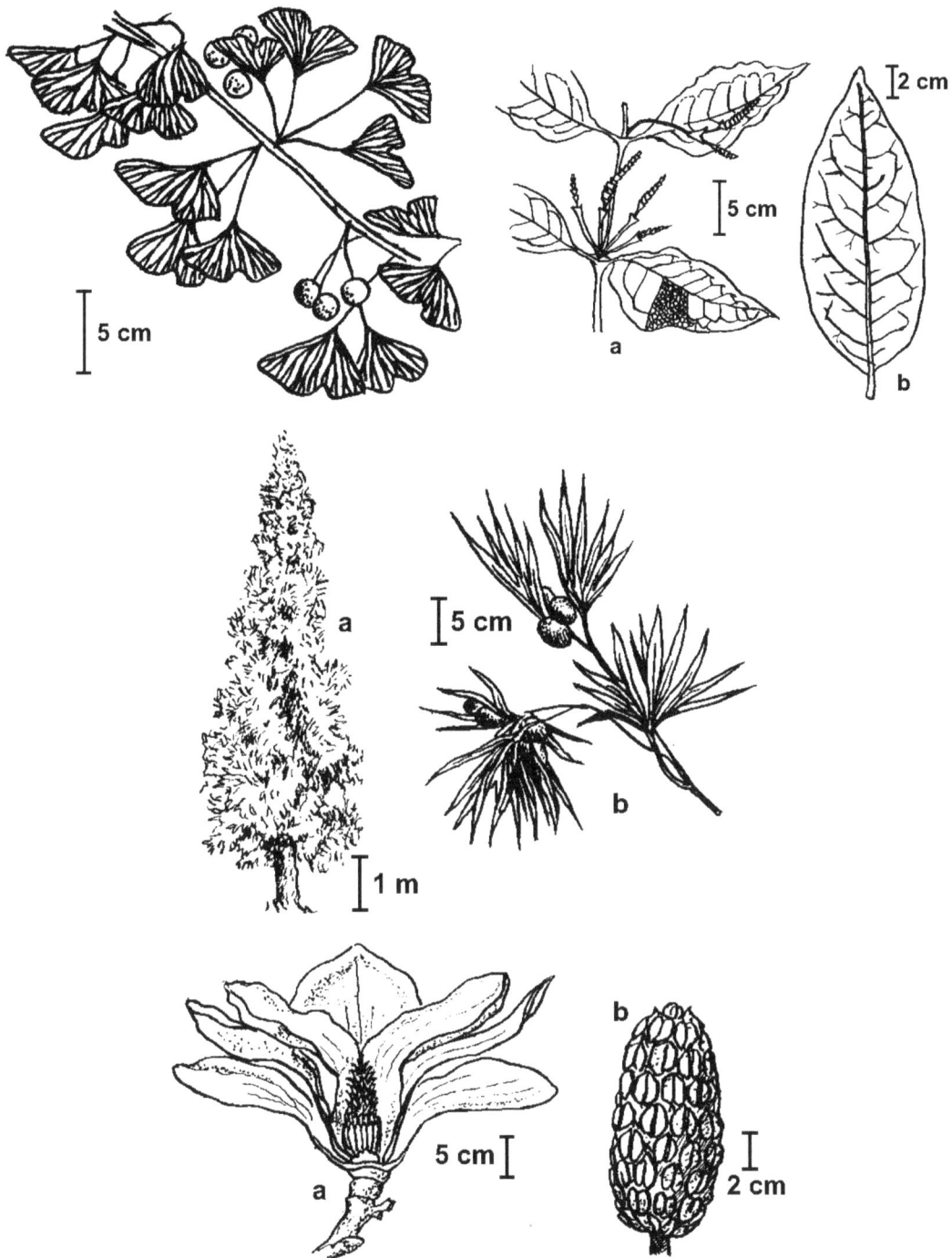

Figure 2.12: **Vegetabilia III (Spermatophyta)**. Left to right, top to bottom: Ginkgoopsida: *Ginkgo biloba*; Gnetopsida: *Gnetum* sp. (a branch with male fructifications, b leaf); Pinopsida: *Podocarpus* sp. (a general view, b brahcn with seeds); Angiospermae: *Magnolia grandiflora* (a flower, b fruit).

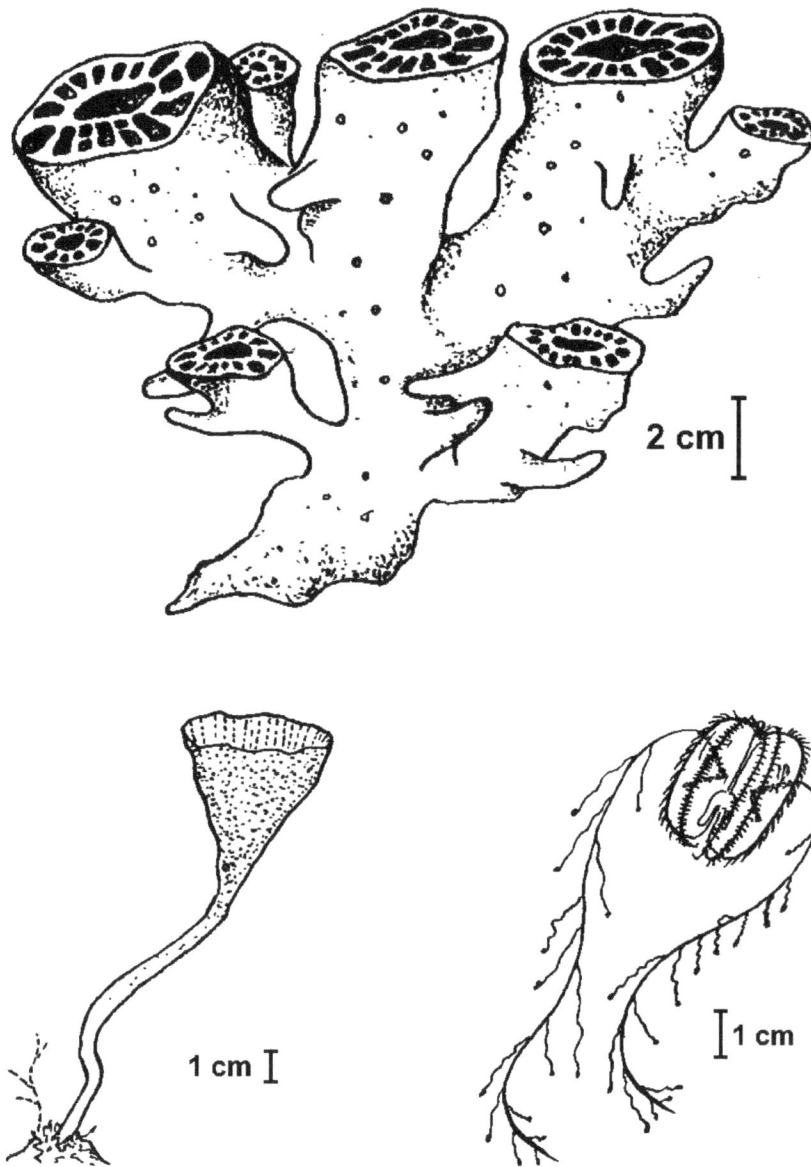

Figure 2.13: **Animalia: Spongia and Ctenophora**. Left to right, top to bottom: Archaeocyatha gen. sp.; Demospongia: *Phakellia cribrosa*; Ctenophora: *Mertensia ovum*.

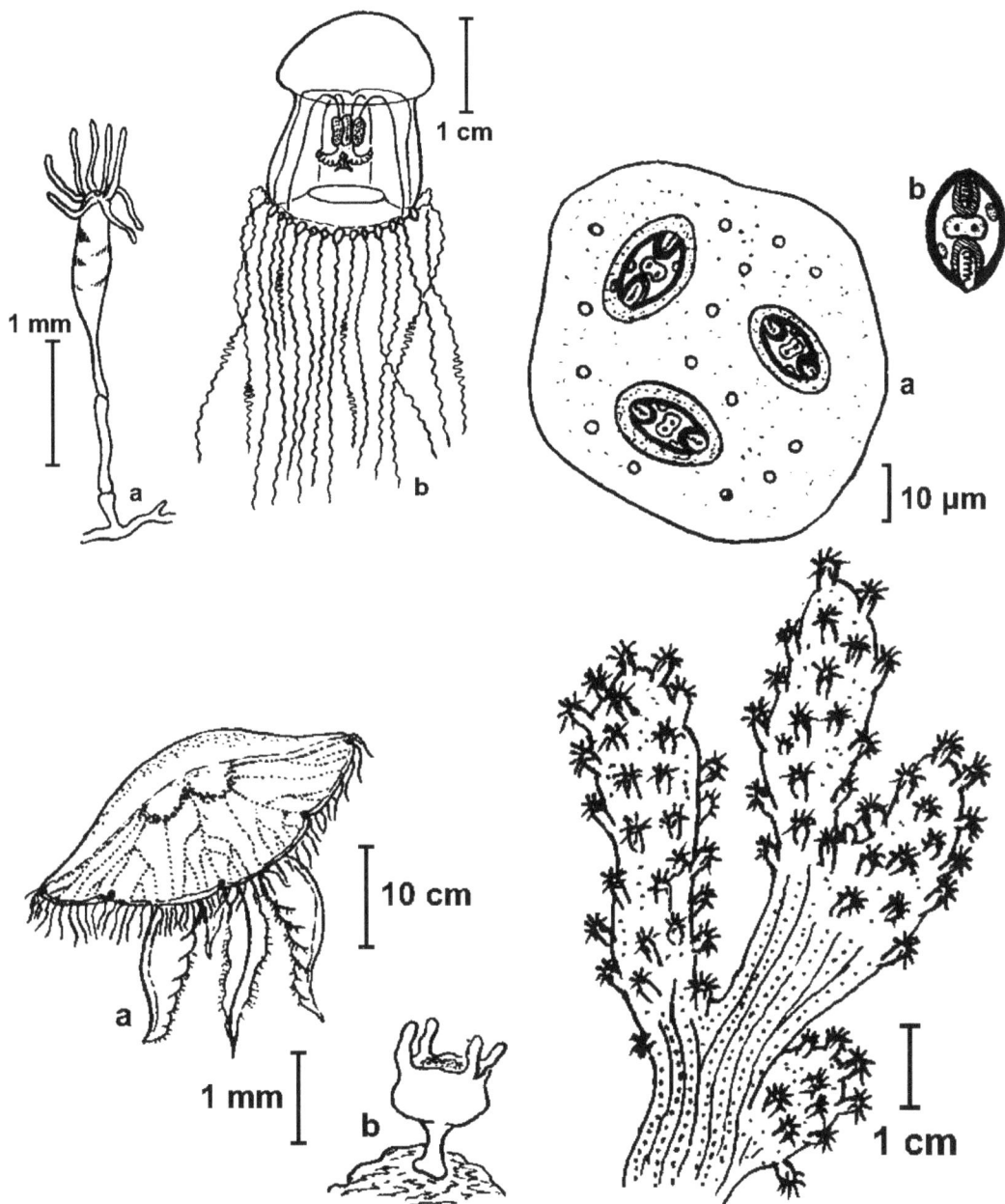

Figure 2.14: **Animalia: Cnidaria**. Left to right, top to bottom: Hydrozoa: *Halitholus yoldiaearcticae* (a polyp stage, b medusa); Myxozoa: *Sphaeromyxa* sp.; Scyphozoa: *Aurelia aurita* (a polyp stage, b medusa); Anthozoa: *Gersemia fruticosa*.

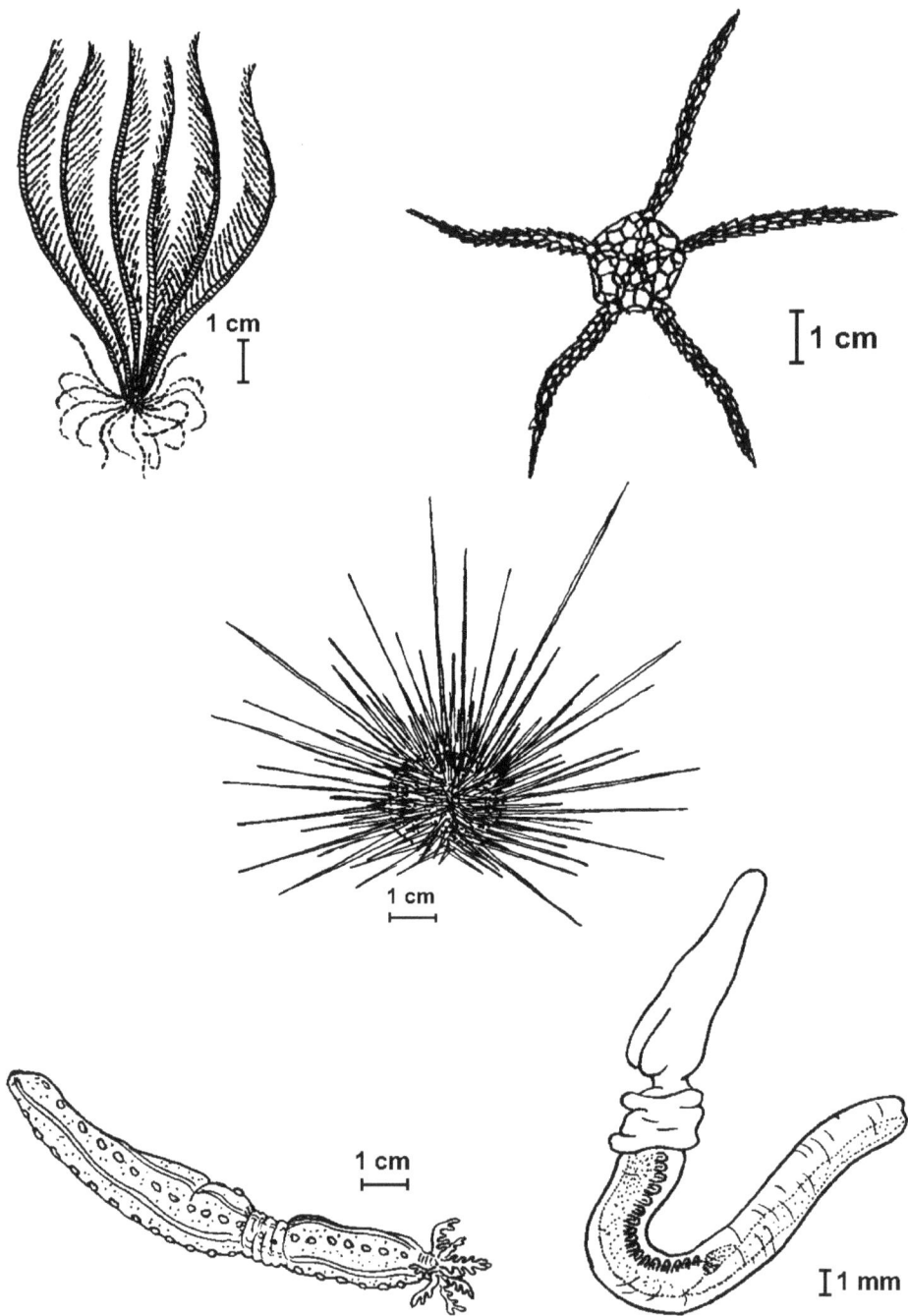

Figure 2.15: **Animalia: Deuterostomia I**. Left to right, top to bottom: Crinoidea: *Heliometra glacialis*; Ophiuroidea: *Stegophiura nodosa*; Echinoidea: *Diadema setosum*; Holothuroidea: *Chiridota laevis*; Enteropneusta: *Saccoglossus mereschkowski*.

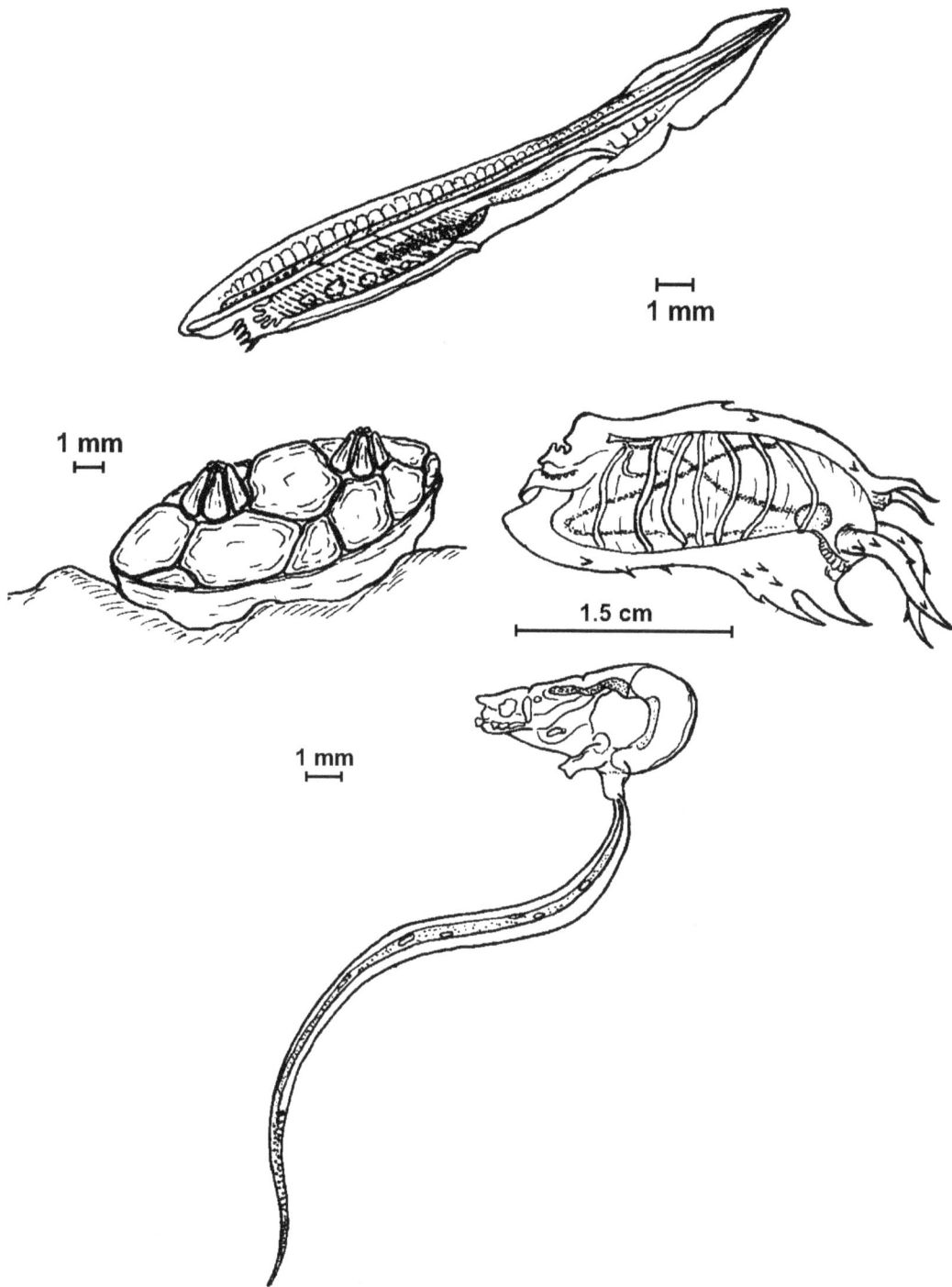

Figure 2.16: **Animalia: Deuterostomia II**. Left to right, top to bottom: Leptocardii: *Branchiostoma lanceolatum*; Ascidiae: *Chelyosoma macleayanum*; Salpae: *Salpa maxima*; Larvacea: *Oikopleura vanhoeffeni*.

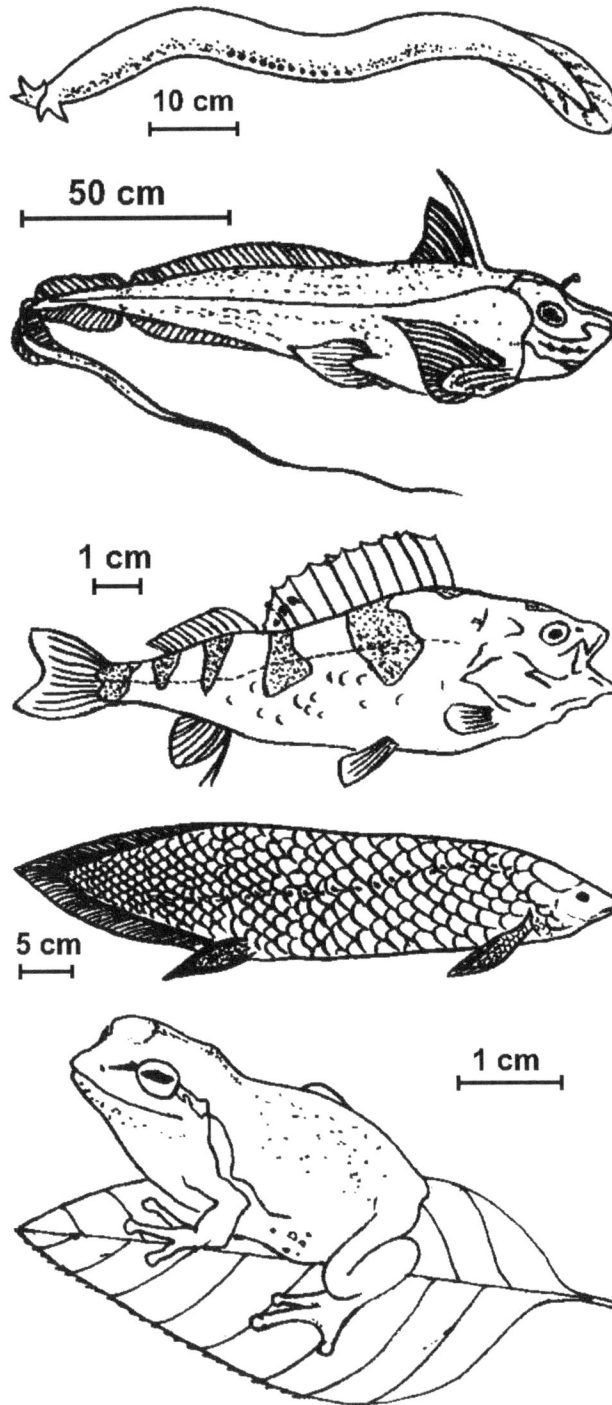

Figure 2.17: **Animalia: Deuterostomia III (Anamnia).** Top to bottom: Cyclostom-ata: *Myxine* sp.; Chondrichtyes: *Chymaera* sp.; Osteichtyes: *Perca fluviatilis*; Dipnoi: *Lepidosiren paradoxa*; Amphibia: *Hyla* sp.

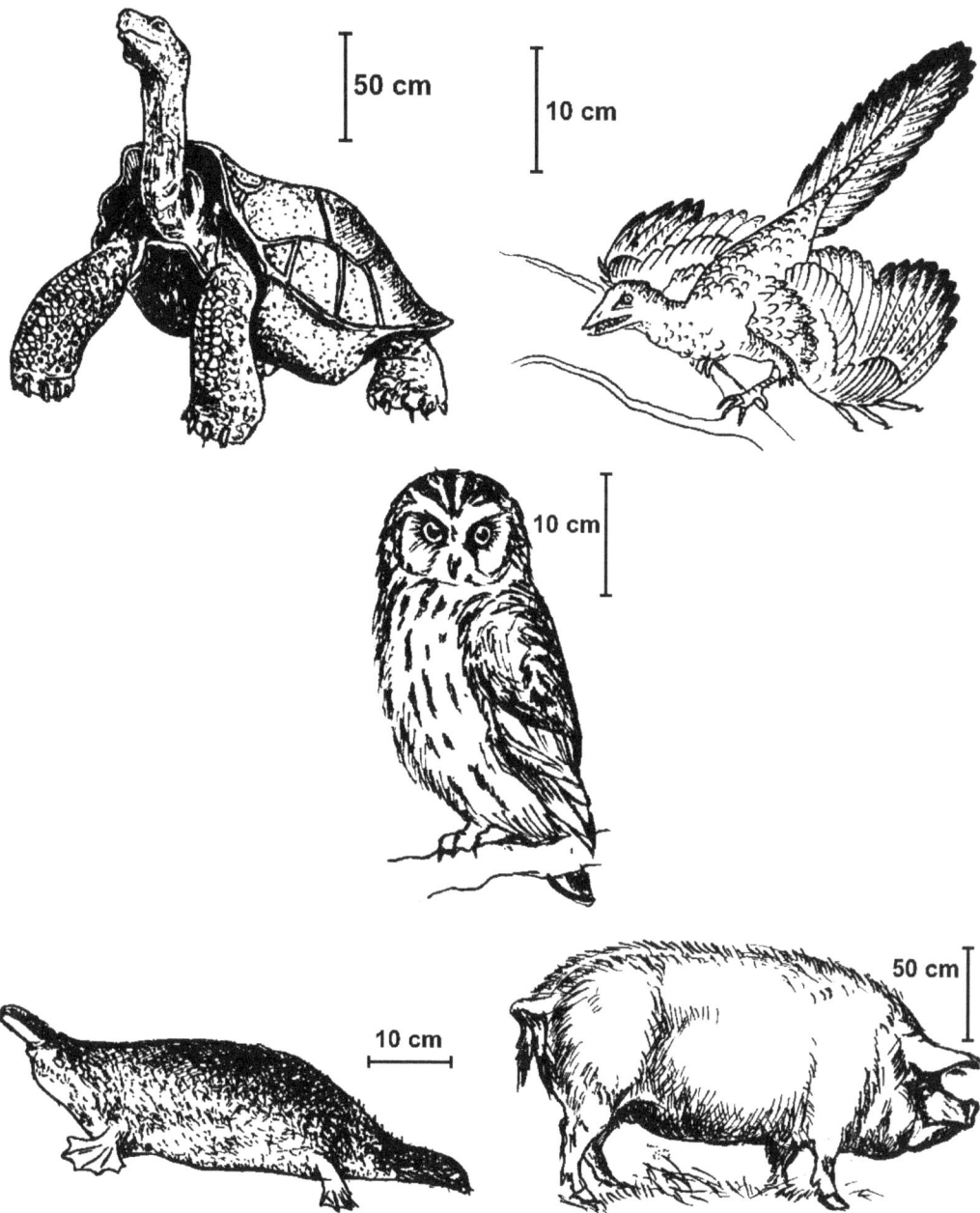

Figure 2.18: **Animalia: Deuterostomia IV (Amniota).** Left to right, top to bottom: Reptilia: *Testudo elephantopus*; Aves: *Archaeopteryx lithographica*; Aves: *Strix aluco*; Mammalia: *Ornithorhynchus anatinus*; Mammalia: *Sus scrofa*.

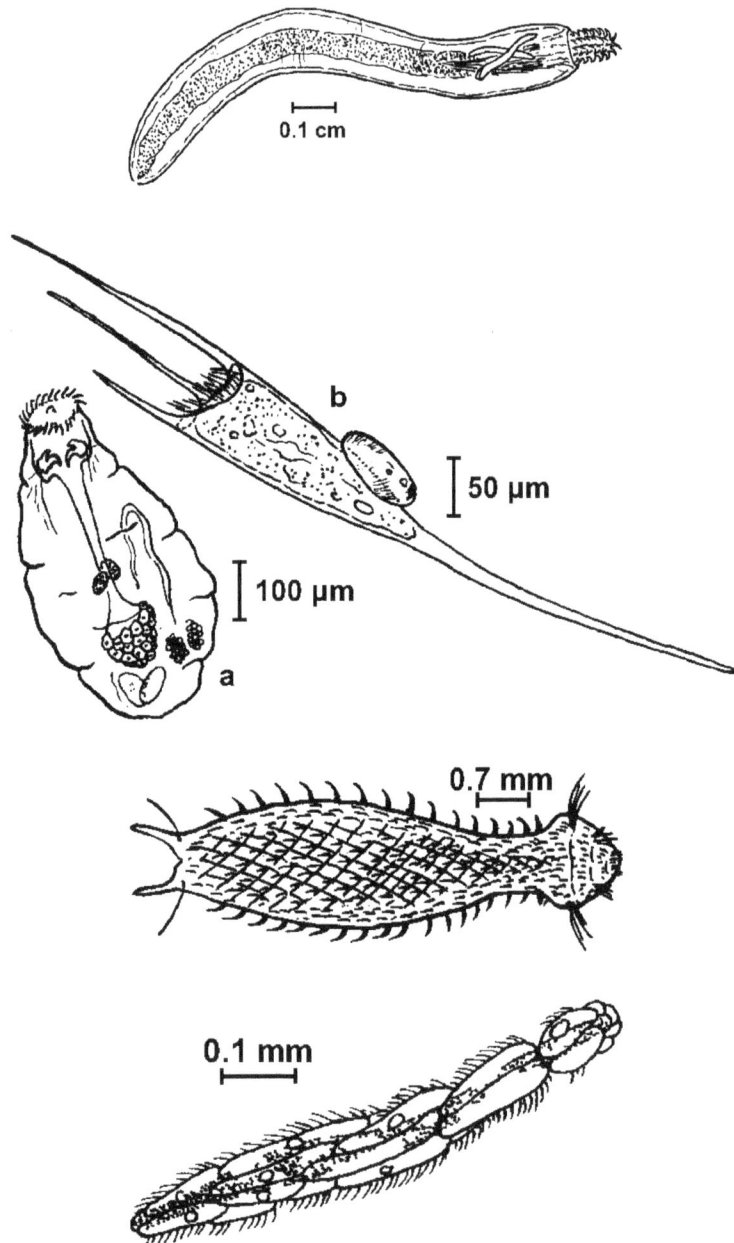

Figure 2.19: **Animalia: Spiralia I**. Left to right, top to bottom: Acanthocephala: *Acanthocephalus lucii*; Rotatoria: a *Asplanchna* sp., b *Kellicottia longispina*; Gastrotricha: *Chaetonotus maximus*; Dicyemea: *Dicyema macrocephalum*.

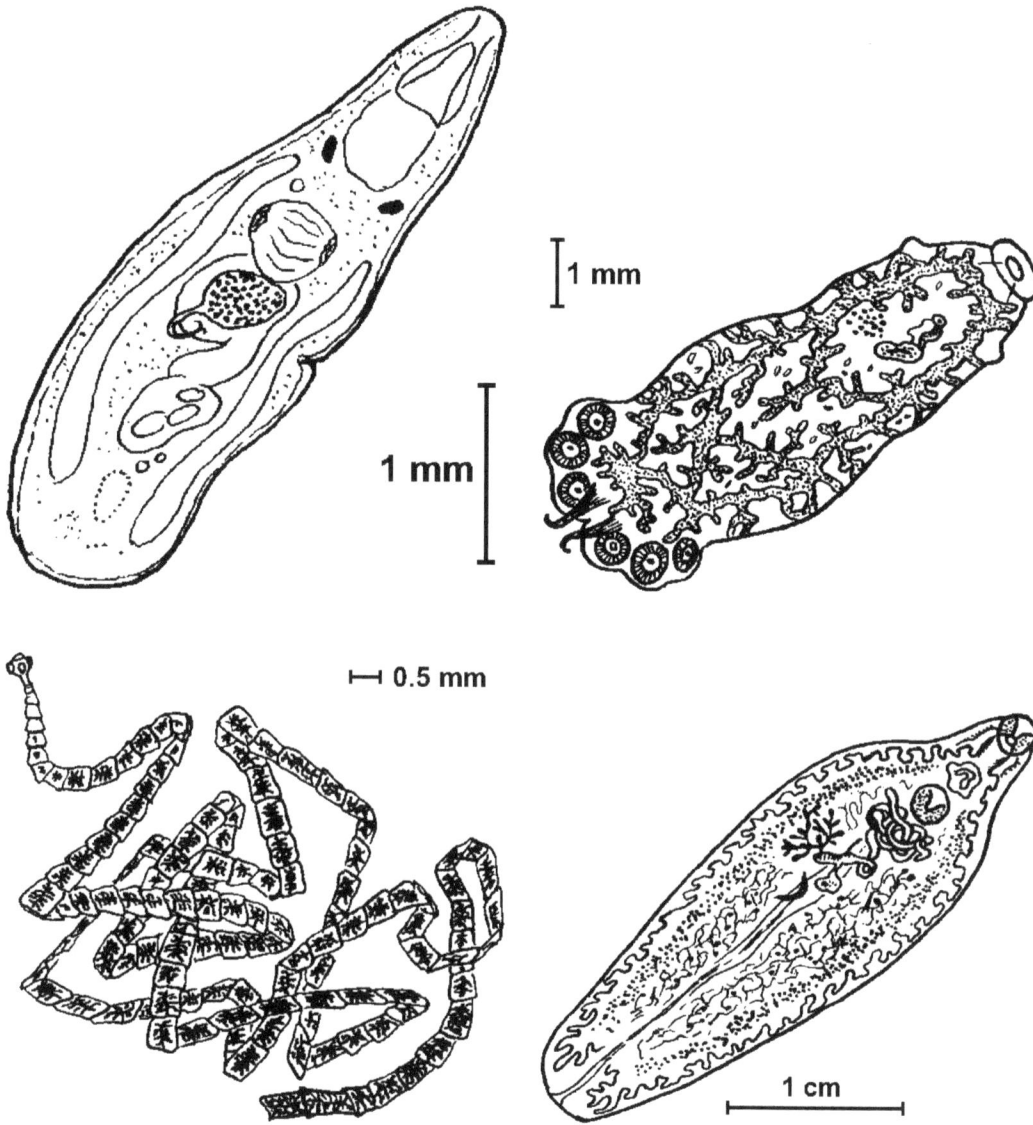

Figure 2.20: **Animalia: Spiralia II (Platyhelminthes)**. Left to right, top to bottom: Rhabditophora: *Macrorhynchus crocea*; Monogenea: *Polystoma integerrimum*; Cestoda: *Taenia solium*; Trematoda: *Fasciola hepatica*.

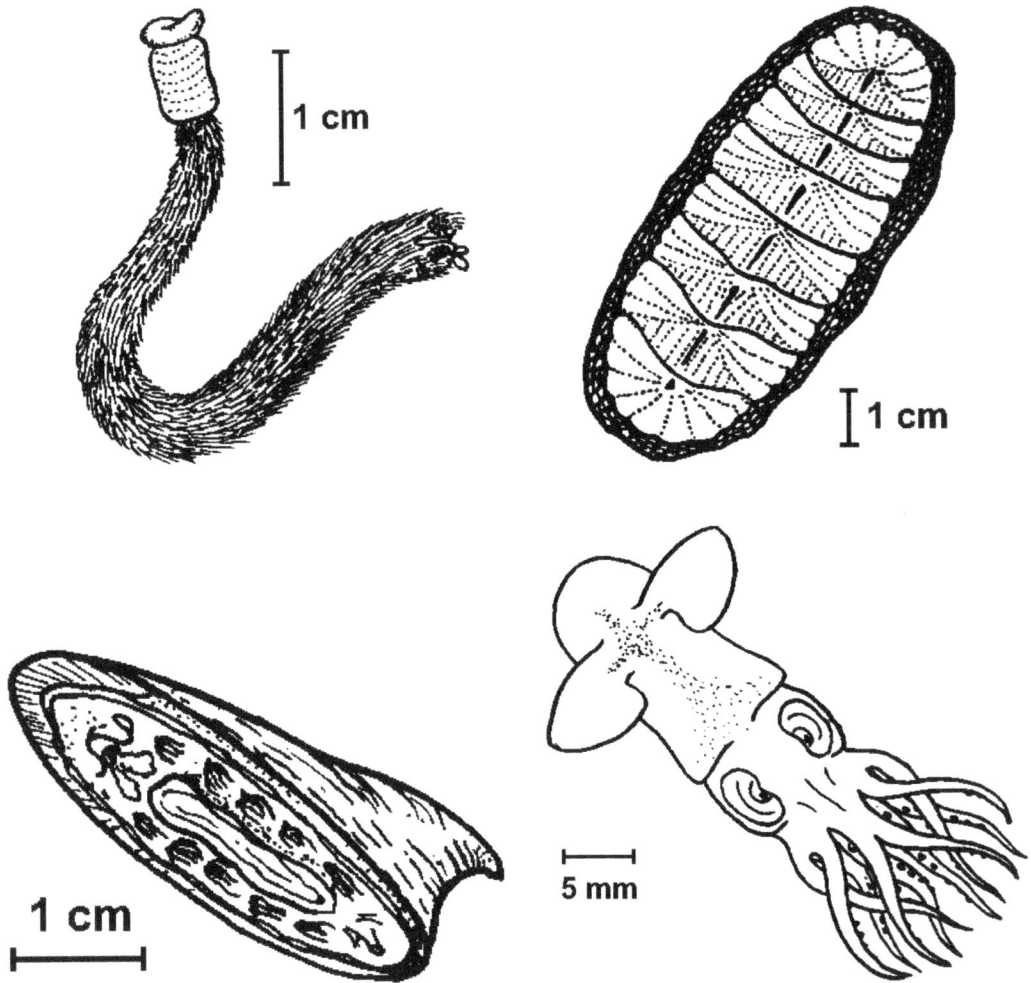

Figure 2.21: **Animalia: Spiralia III (Mollusca I)**. Left to right, top to bottom:
Aplacophora: *Chaetoderma nitidulum*; Polyplacophora: *Chiton sulcatus*; Monopla-
cophora: *Neopilina galatheae*; Cephalopoda: *Sepiola birostrata*.

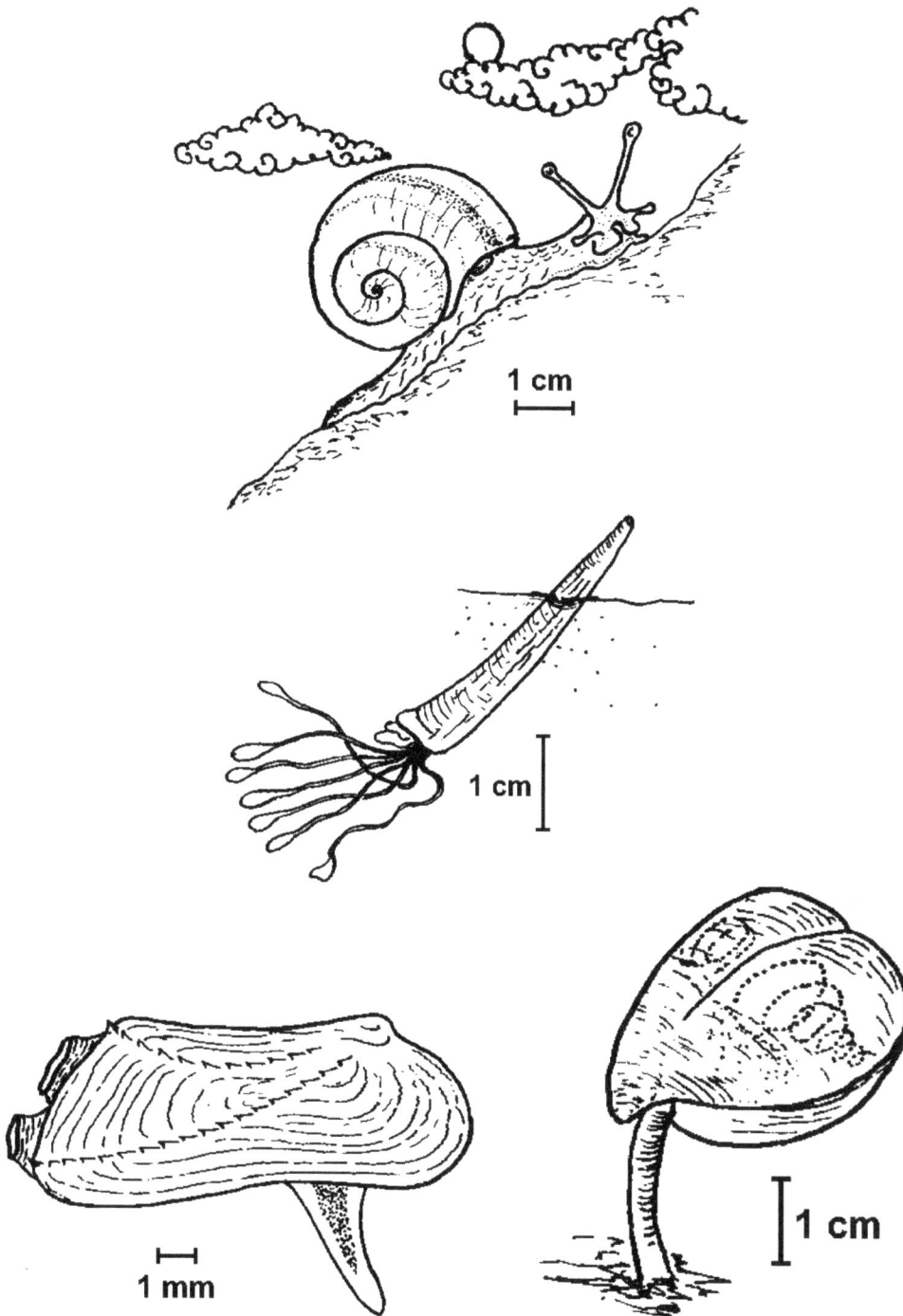

Figure 2.22: **Animalia: Spiralia IV**. Left to right, top to bottom: Gastropoda: *Helix vulgaris*; Scaphopoda: *Antalis vulgaris*; Bivalvia: *Hiatella arctica*; Brachiopoda: *Spirifer* sp. (fossil).

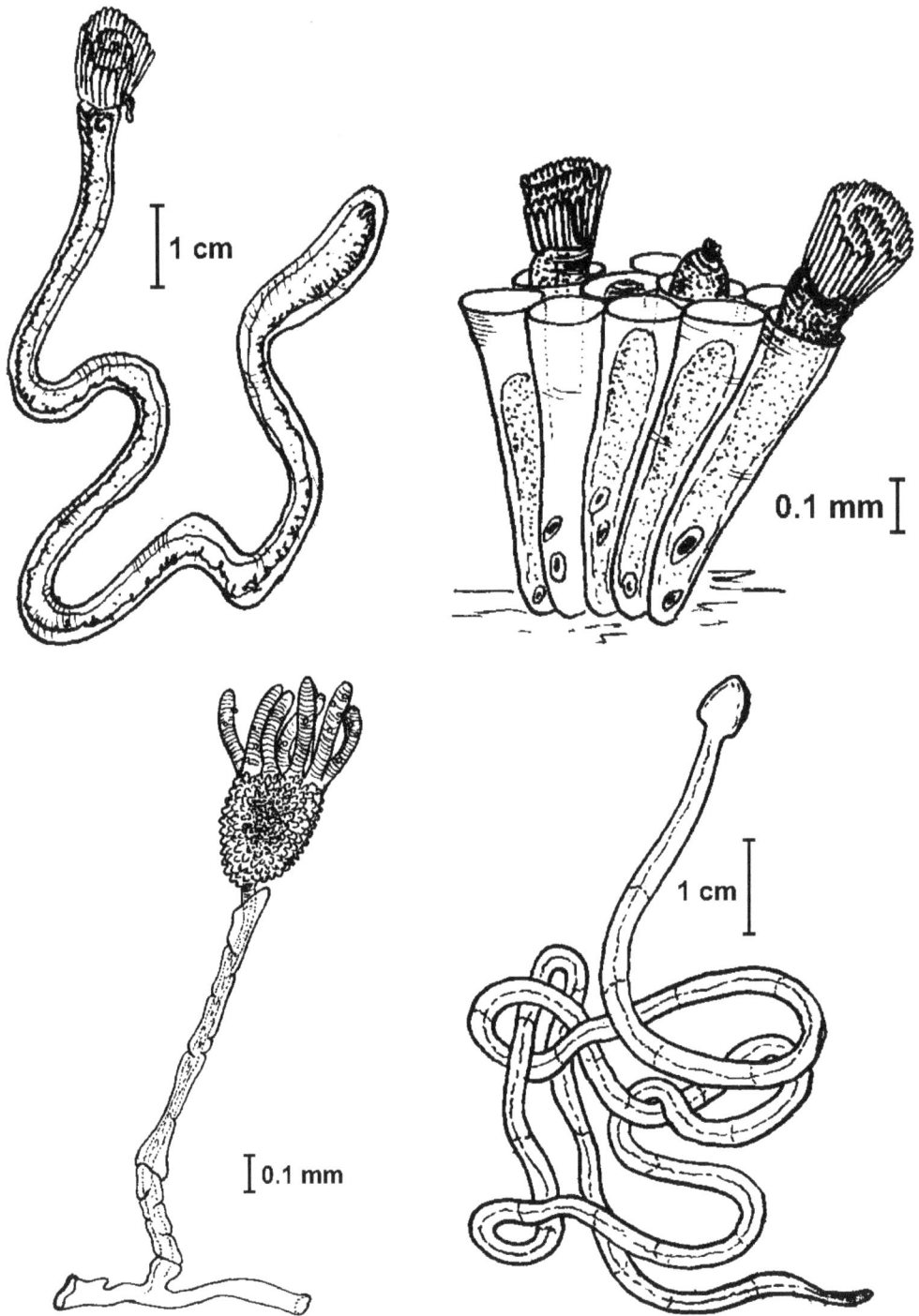

Figure 2.23: **Animalia: Spiralia V**. Left to right, top to bottom: Phoronida: *Phoronis hippocrepia*; Bryozoa: *Plumatella fungosa*; Entoprocta: *Pedicellina nutans*; Nemertea: *Emplectonema* sp.

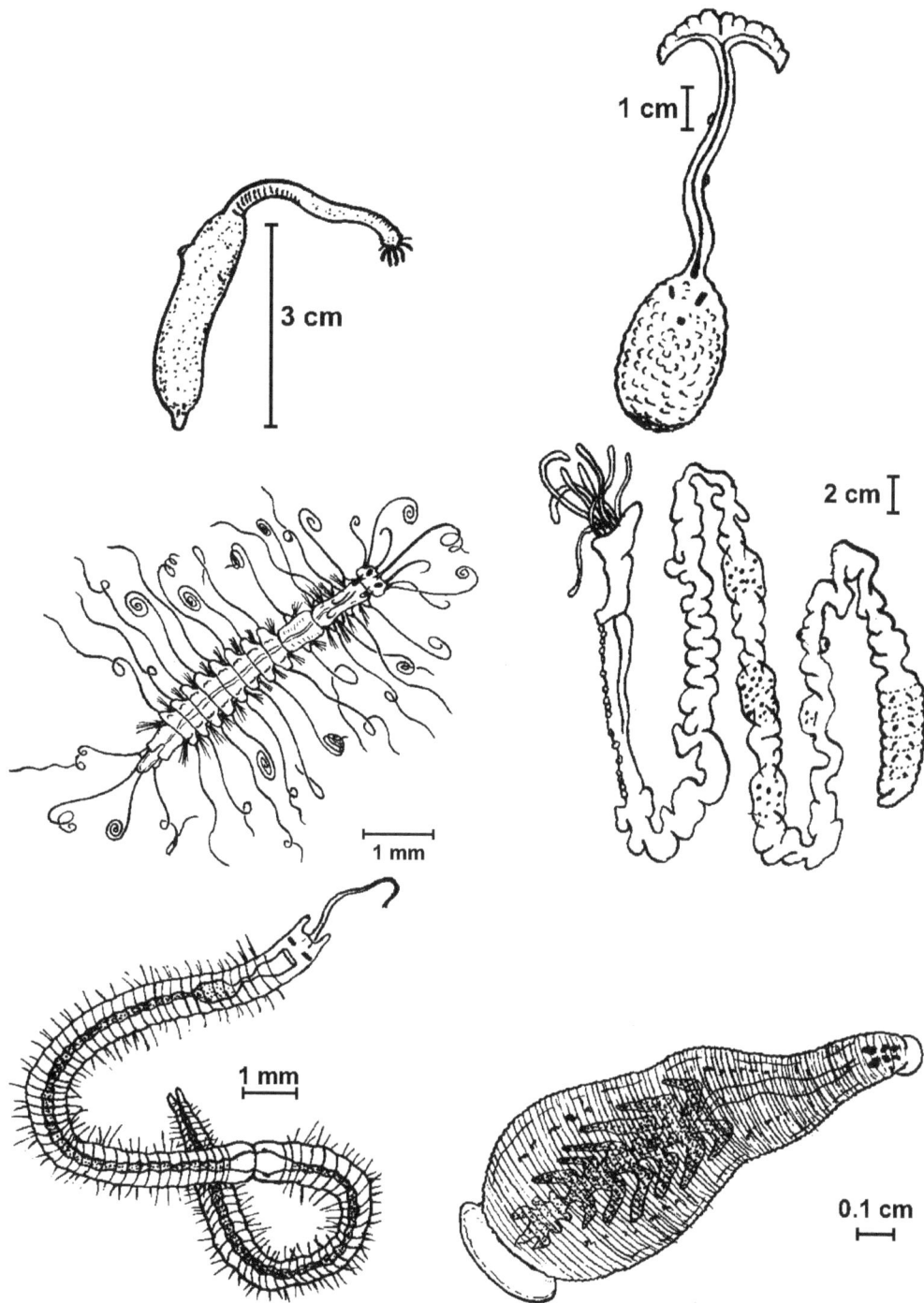

Figure 2.24: **Animalia: Spiralia VI**. Left to right, top to bottom: Sipuncula: *Golfingia margaritacea*; Echiura: *Bonellia viridis*; Polychaeta: *Pterosyllis finmarchica*; Pogonophora: *Choanophorus indicus*; Clitellata: *Stylaria lacustris*; Hirudinea: *Glossiphonia complanata*.

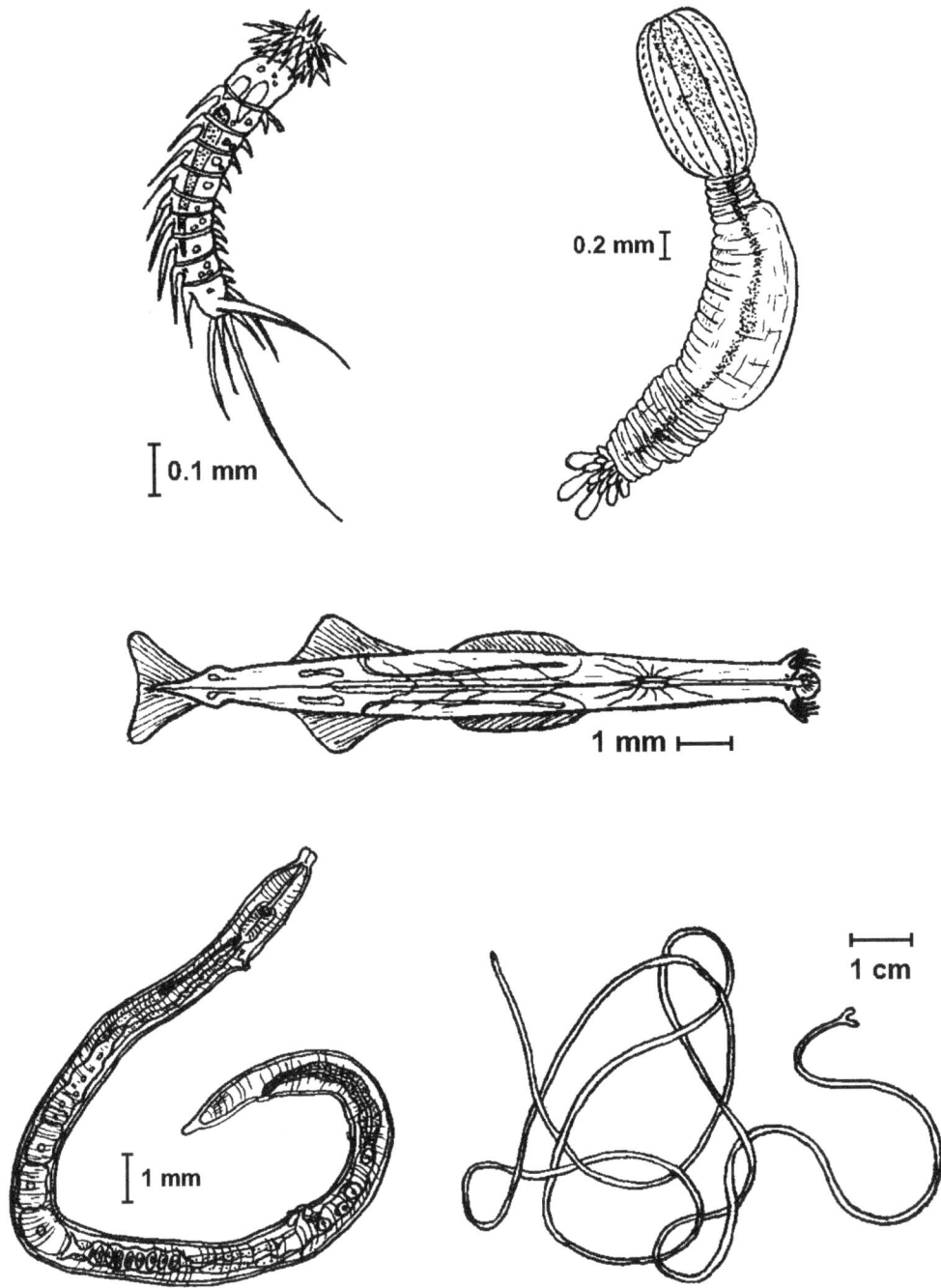

Figure 2.25: **Animalia: Chaetognatha and Ecdysozoa I**. Left to right, top to bottom: Chaetognatha: *Sagitta hexaptera*; Priapulida: *Priapulus caudatus*; Kynorhyncha: *Semnoderis armiger*; Nematoda: *Aphelenchoides composticola*; Nematomorpha: *Gordius aquaticus*.

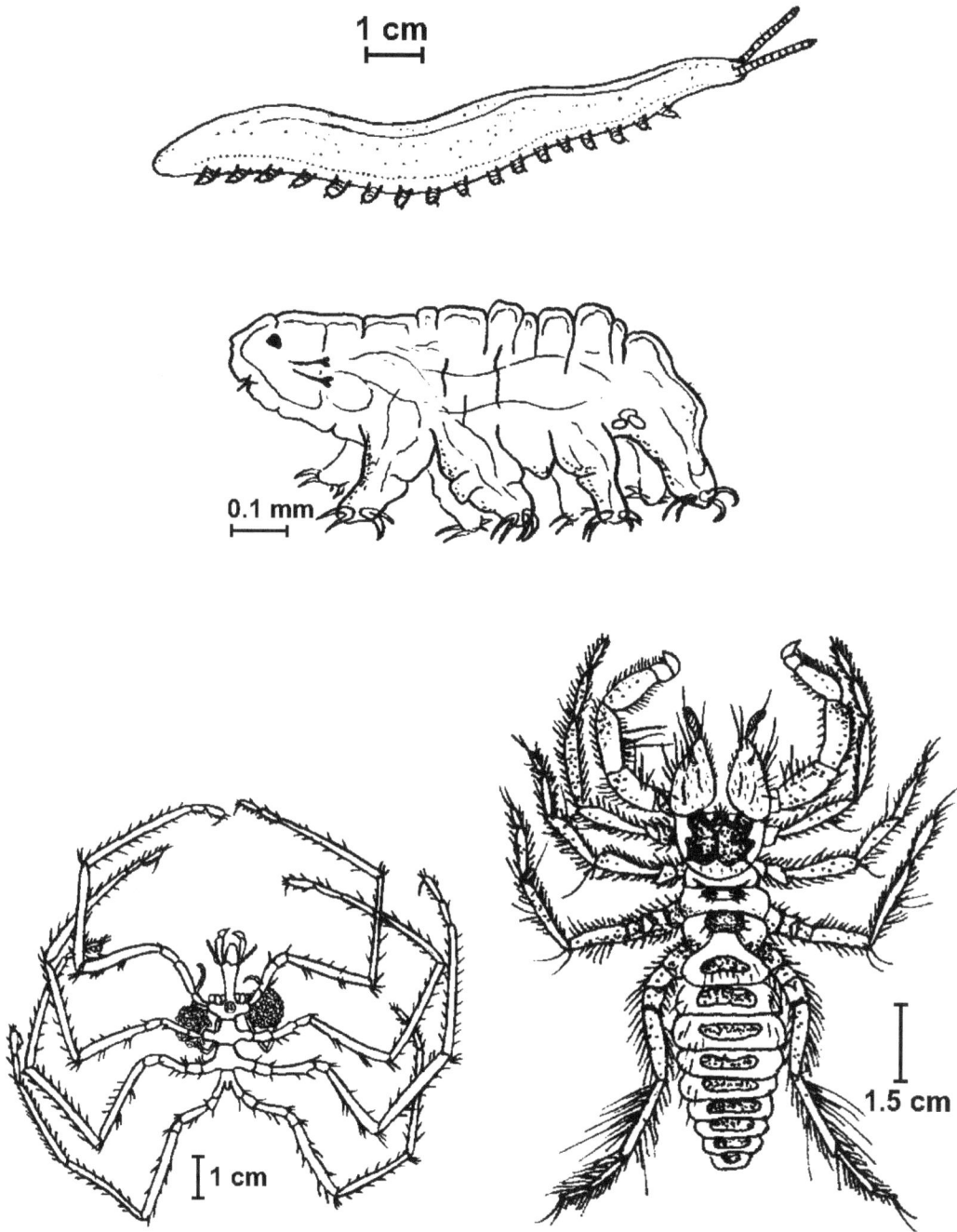

Figure 2.26: **Animalia: Ecdysozoa II**. Left to right, top to bottom: Onychophora: *Peripatopsis capensis*; Tardigrada: *Macrobiotus* sp.; Pantopoda: *Nymphon brevirostre*; Chelicerata: *Galeodes araneoides*.

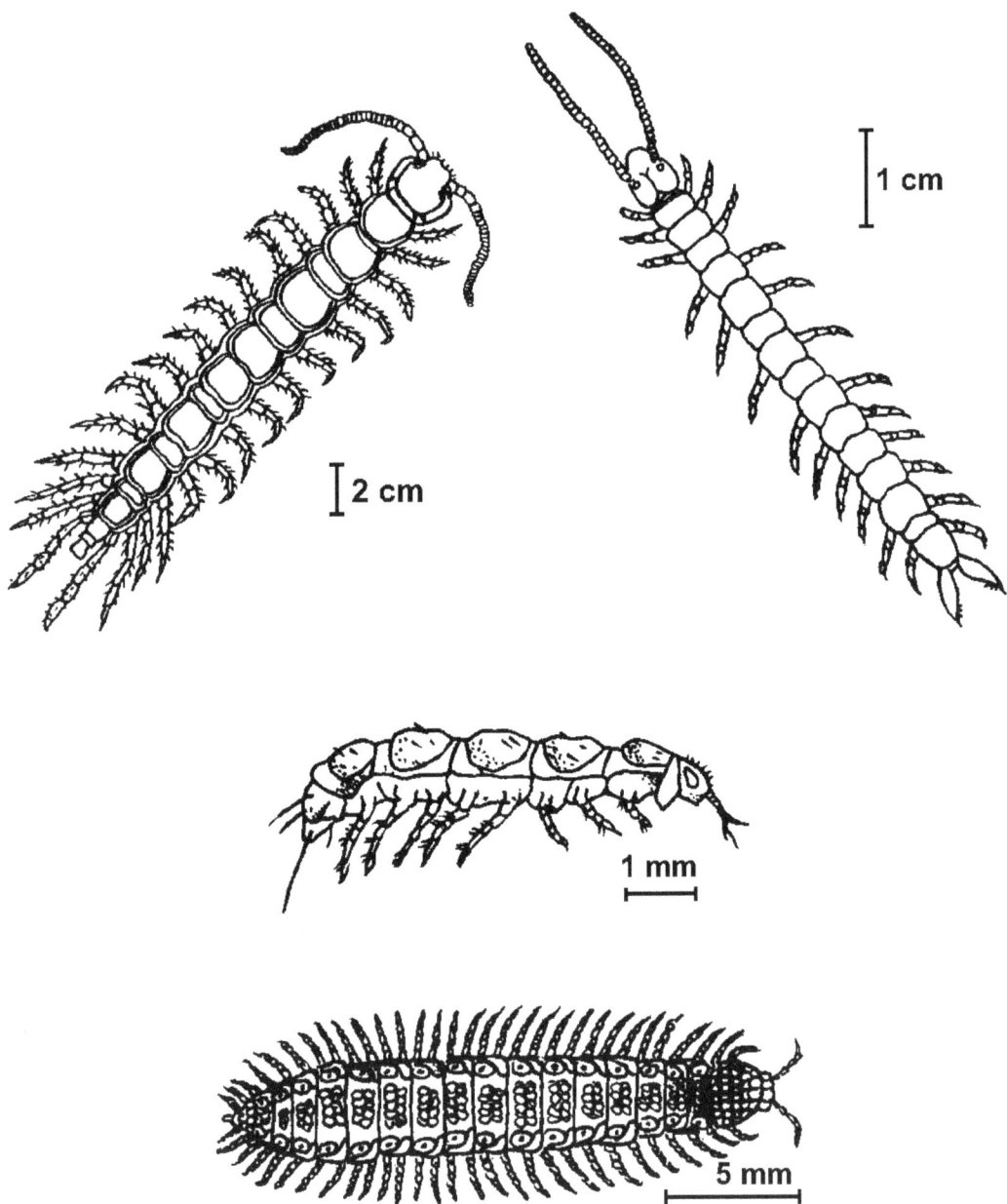

Figure 2.27: **Animalia: Ecdysozoa III (Myriapoda)**. Left to right, top to bottom: Chilopoda: *Lithobius forficatus*; Symphyla: *Scutigerella immaculata*; Pauropoda: *Pauropus sylvaticus*; Diplopoda: *Polydesmus complanatus*.

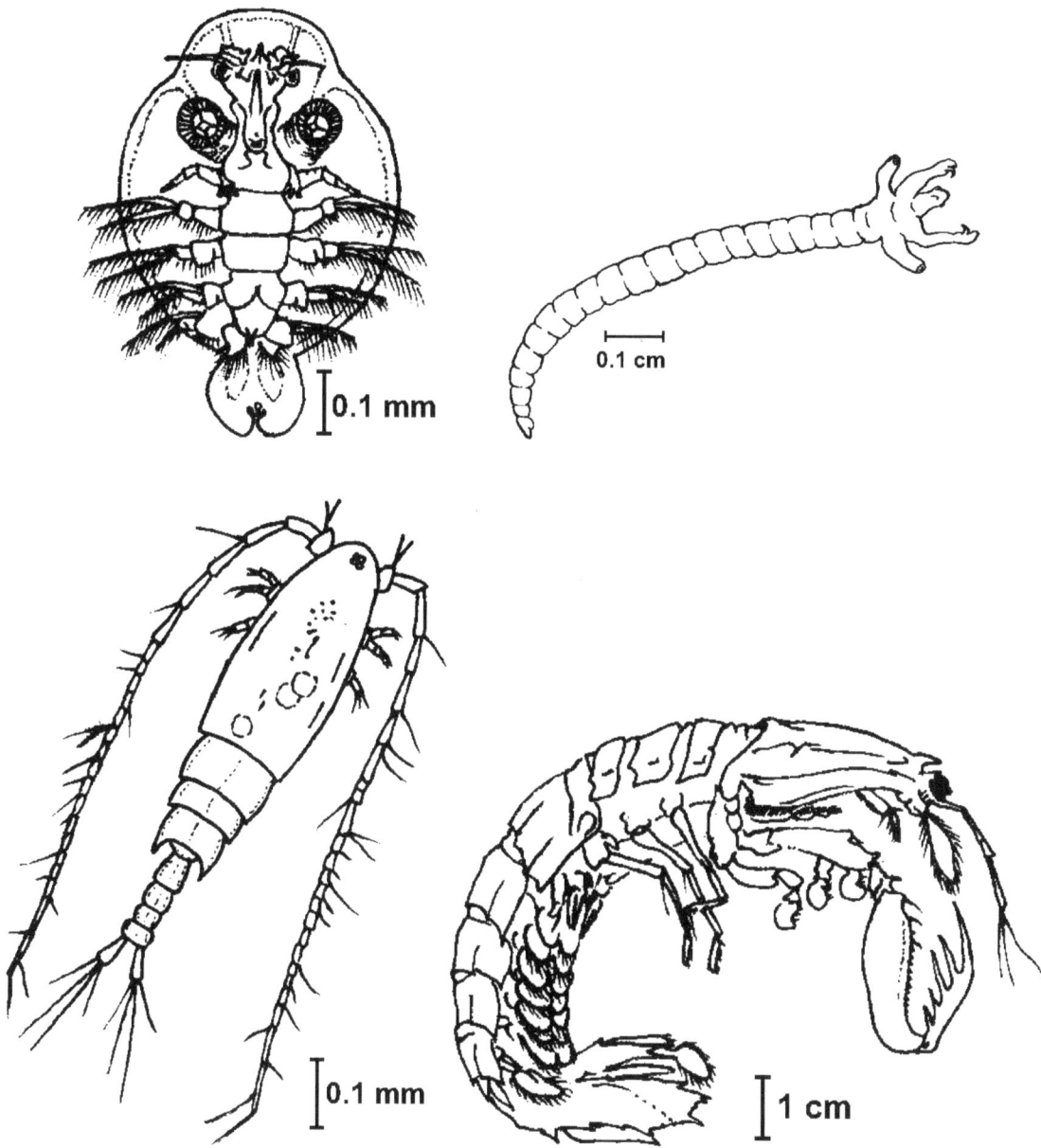

Figure 2.28: **Animalia: Ecdysozoa IV (Pancrustacea I)**. Left to right, top to bottom: Branchiura: *Argulus* sp.; Pentastomida: *Cephalobaena tetrapoda*; Copepodoidea: *Calanus* sp.; Malacostraca: *Squilla mantis*.

8

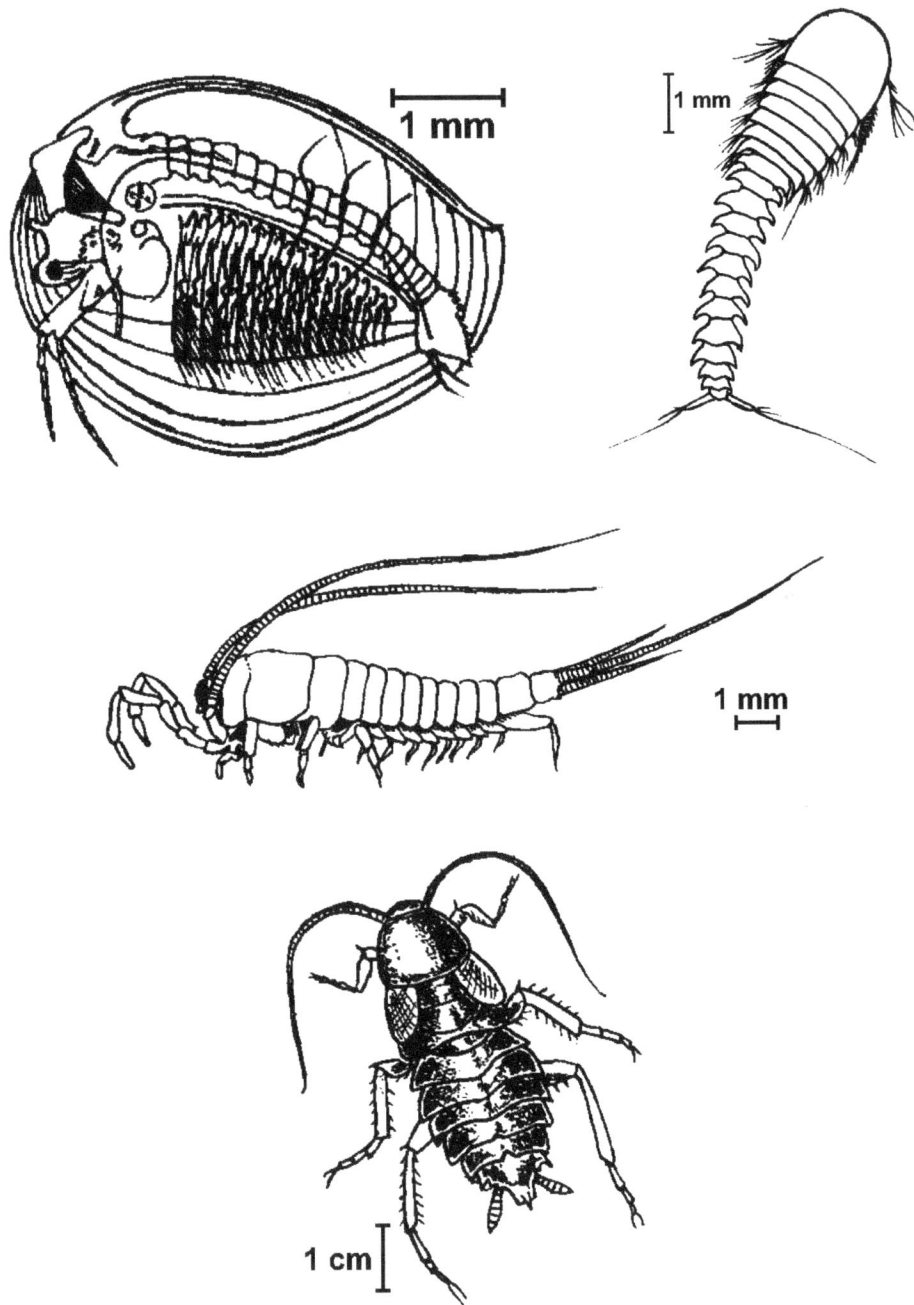

Figure 2.29: **Animalia: Ecdysozoa V (Pancrustacea II)**. Left to right, top to bottom: Phyllopoda: *Limnadia lenticularis*; Cephalocarida: *Sandersiella acuminata*; Archaeognatha: *Machilis* sp.; Hexapoda: *Blatta orientalis*.

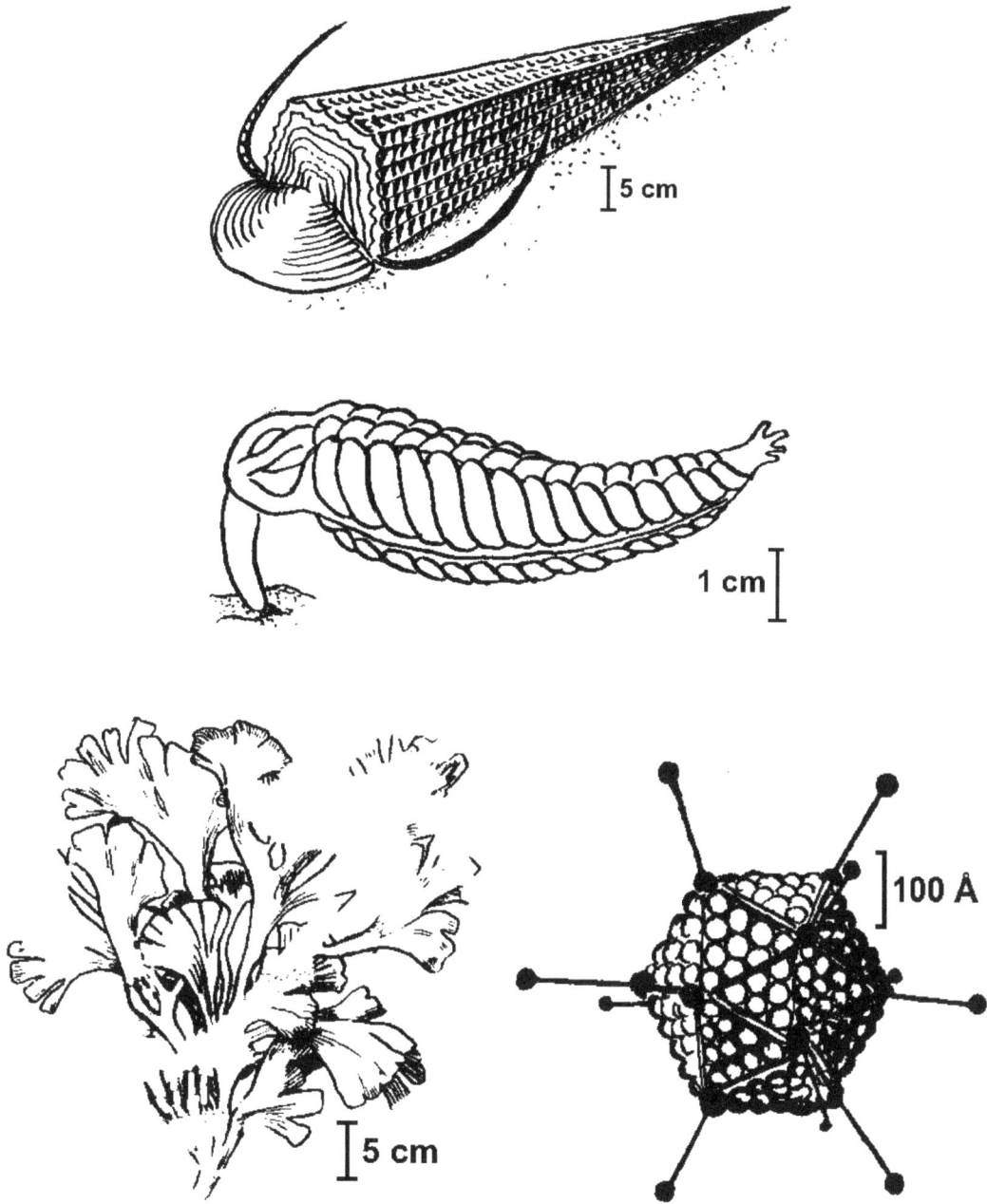

Figure 2.30: **Problematica** and **Viri**. Left to right, top to bottom: Hyolitha gen. sp.; Erniettiomorpha: *Pteridinium* sp.; "Nematophyta": *Prototaxites* sp.; Viri: *Adenovirus* sp.

Chapter 3

Life Stories

3.1 Foundations of Science

Principle of Actuality

It states that current geologic processes, occurring at the same rates observed today, in the same manner, account for all of Earth's geological features. The central argument of this principle is "The present is the key to the past."

Occam's razor

The principle of simplicity is the central theme of father William of Occam's (ca. 1300) approach, so much so that this principle has come to be known as "Occam's Razor." Occam used his principle to eliminate unnecessary hypotheses and favor shortest explanation (which nowadays is called most parsimonous).

Principle of Falsification

The principle of falsification by sir Karl Popper defined as "A theory is falsifiable ... if there exists at least one ... statement[s] which [is] forbidden by it". An example of non-falsifiable hypothesis could be the Russel's teapot: "If I were to suggest that between the Earth and Mars there is a china teapot revolving about the sun in an elliptical orbit, nobody would be able to disprove my assertion provided I were careful to add that the teapot is too small to be revealed even by our most powerful telescopes."

Null and Alternative Hypothesis

Ronald Fisher, British statistician and geneticist formulated an approach of two rival hypotheses: null hypothesis and the alternative hypothesis. If being tested with alternative hypothesis, it is failed-to-reject if the null hypothesis is rejected based on statistical evidence.

3.2 Few Drops of Geology

Geological Time

Stratigraphy refers to the natural and cultural soil layers that make up an archaeological deposit. The notion is tied up with 19th-century geologist Charles Lyell, who stated that because of natural forces, soils found deeply buried will have been laid down earlier—and therefore be older—than the soils found on top. The age of rock layers are determined by fossils using **relative dating** and **absolute dating**. If there are same fossils in more than one layers, these layers have the same age The majority of the time fossils are dated using relative dating techniques. Using relative dating the fossil is compared to something for which an age is already known.

Absolute dating is used to determine a precise age of a rock or fossil through radiometric dating methods. Radioactive minerals occur in rocks and fossils are almost a geological clock. Radioactive isotopes break down at a constant rate over time through radioactive decay. By measuring the ratio of the amount of the stable isotope to the amount of the radioactive isotope, an age can be determined.

Origin of Earth

Pierre-Simone Laplace reached the conclusion that the stability of the Solar system would best be accounted for by a process of evolving from chaos. Laplace suggested that:

1. The Sun was originally a giant cloud of gas or nebulae that rotated evenly.

2. The gas contracted due to cooling and gravity.

3. This forced the gas to rotate faster, just as an ice skater rotates faster when his extended arms are drawn onto his chest.

4. This faster rotation would throw off a rim of gas, which following cooling, would condense into a planet.

5. This process would he repeated several times to produce all the planets.

6. The asteroids between Mars and Jupiter were caused by rings which failed to condense properly.

7. The remaining gas ball left in the centre became the Sun.

It is frequently accepted nowadays that in addition to the above processes, Earth underwent the heating stage and at some point likely became a "lava ball", and then cooling stage when water start to condense and make primary ocean. Also, geology and astronomical features of Moon suggest that this body *originated from Earth* on the some very early stage of Solar system evolution.

Structure of Earth

The Earth consists of concentric layers, core, mantle and crust. **Core** is in the center and is the hottest part of the Earth. It is solid and possibly made up of metals with temperatures of up to 5,500°C. **Mantle** is the widest section of the Earth. It has a thickness of approximately 2,900 km. The mantle is made up of semi-liquid magma. **Crust** is the outer layer of the Earth. It is a thin layer up to 60 km deep. The crust is made up of tectonic plates, which are in constant motion. Earthquakes and volcanoes are most likely to occur at plate boundaries.

Everything in Earth can be placed into one of four major subsystems: lithosphere (land), hydrosphere (water), biosphere (living things), and atmosphere (air). Earth is the only known planet that has a layer of water.

The differentiation of Earth body finally resulted in developing of lighter gas layer on the surface (primary atmosphere), initially very thin and relatively cold. Therefore, water vapour were condensed into primary ocean (primary hydrosphere). According to the principle of actuality, it should be close to today's volcanic gases 15% of CO_2, plus CH_4 (methane), NH_3 (ammonia), H_2S, SO_2 and different "acidic smokes" like HCl.

Plate Tectonics

Alfred Wegener is best known as the creator of the theory of continental drift by hypothesizing in 1912 that the continents are slowly gliding around the Earth. According to Wegener, in the beginning of Mesozoic era, there were two large continents, Gondwana and Laurasia which were separated by Tethys Ocean. Gondwana was one of Earth broke up around 180 000 years ago. Moreover, in Permian period, all continents where united in one as Pangaea, which was surrounded by one big ocean.

Mantle is the thickest Earth layer but it slowly moves. Mantle convection breaks the lithosphere into plates and continues to move them around Earth surface. These plates might move alongside each other, move by, and even collide with each other.

As a result, oceans basins may open, it may move continents, create mountains, and cause earthquakes. Continents will keep changing their positions due to mantle convection.

Hotspots are the living proofs of mantle convection. In the United States, there are two locations which are considered hotspots: Yellowstone and Hawaii. A hotspot is a place where the intense heat of the outer core radiates through the mantle. The most amazing fact about them is that whereas ocean (Hawaii) or continental (Yellowstone) plates move on, these hotspots stay in place! This is why in the past, Yellowstone was located westward, and Hawaiian volcanoes northward.

3.3 Chemistry of Life

To understand life, a basic knowledge of chemistry is needed. This includes atoms (and its components like protons, neutrons and electrons), atomic weight, isotopes, elements, the periodic table, chemical bonds (ionic, covalent, and hydrogen), valence, molecules, and molecular weight:

Atoms The smallest unit of matter undividable by chemical means. An atom is made up of two main parts: a nucleus vibrating in the centre, and a virtual cloud of electrons spinning around in zones at different distances from the nucleus (not to be confused with the cell nuclei). When atoms interact, its less like the bumping of balls and more a matter of attraction and repulsion; at atomic levels, mass is much less an important consideration than charges, which are electrical: positive, negative, or neutral (balanced).

Protons Stable elementary particles having the smallest known positive charge, found in the nuclei of all elements. The proton mass is less than that of a neutron. A proton is the nucleus of the light hydrogen atom, i.e., the hydrogen ion.

Neutrons Like protons but neutral charge and also located in the nucleus.

Radioactive decay occurs in unstable atomic nuclei—that is, ones that do not have enough binding energy to hold the nucleus together due to an excess of either protons or neutrons.

Electrons determine how atoms will interact. Located on the outside of the nucleus known as the outer shell and has a negative charge that determines what type of change it has.

Atomic weight is the average of the masses of naturally-occurring isotopes.

67

Isotopes each of two or more forms of the same element that contain equal numbers of protons but different numbers of neutrons.

Periodic table of elements The periodic table we use today is based on the one devised and published by Dmitri Mendeleev in 1869. The periodic table of the chemical elements displays the organization of matter.

Chemical bonds are attractive forces between the atoms.

Valence A typical number of chemical bonds in this element. For example, nitrogen (N) has valence 3 and therefore usually makes three bonds.

Molecules Molecules form when two or more atoms form chemical bonds with each other.

Molecular weight calculated as a sum of atomic weights.

Covalent bonds When two atoms bind in one molecule, there are two variants possible. Non-polar bond is when electrons are equally shared between atoms. Polar bond is when one of atoms attract electrons more than another, and therefore becomes partly negative while the second—partly positive.

Ionic bonds Ionic bonding is the complete transfer of some electron(s) between atoms and is a type of chemical bond that generates two oppositely charged ions.

It is essential to know that protons have a charge of $+1$, neutrons have no charge, and electrons have a charge of -1. The atomic weight is equal to the weight of protons and neutrons. Isotopes have the same number of protons but different number of neutrons; some isotopes are unstable (radioactive).

One of the most outstanding molecules is water. Theoretically, water should boil at much lower temperature, but it boils at $100°C$ just because of the hydrogen bonds sealing water molecules. These bonds arise because a water molecule is *polar*: hydrogens are slightly positively charged, and oxygen is slightly negatively charged.

Another important concept related to water is **acidity**. If in a solution of water, the molecule takes out proton (H^+), it is an **acid**. One example of this would be hydrochloric acid (HCl) which dissociates into H^+ and Cl^-. If the molecule takes out OH^- (hydroxide ion), this is a **base**. An example of this would be sodium hydroxide (NaOH) which dissociates into Na^+ and hydroxide ion.

To plan chemical reactions properly, we need to know about **molar mass** and **molar concentration**. Molar mass is a gram equivalent of molecular mass. This means that (for example) the molecular mass of salt (NaCl) is $23 + 35$, which equals 58.

Consequently, one mole of salt is 58 grams. One mole of any matter (of molecular structure) always contains $6.02214078 \times 10^{23}$ molecules (**Avogadro's number**).

The density of a dissolved substance is the **concentration**. If in 1 liter of distilled water, 58 grams of salt are diluted, we have 1M (one molar) concentration of salt. Concentration will not change if we take any amount of this liquid (spoon, drop, or half liter).

Depending on the concentration of protons in a substance, a solution can be very acidic. The acidity of a solution can be determined via pH. For example, if the concentration of protons is 0.1 M (1×10^{-1}, which 0.1 grams of protons in 1 liter of water), this is an extremely acidic solution. The pH of it is just 1 (the negative logarithm, or negative degree of ten of protons concentration). Another example is distilled water. The concentration of protons there equals 1×10^{-7} M, and therefore pH of distilled water is 7. Distilled water is much less acidic because water molecules dissociate rarely.

When two or more carbon atoms are connected, they form a **carbon skeleton**. All **organic molecules** are made of some organic skeleton. Apart from C, elements participate in organic molecules (biogenic elements) are H, O, N, P, and S. These six elements make four types of biomolecules: (1) lipids—hydrophobic organic molecules which do not easily dissolve in water; (2) carbohydrates or sugars, such as glucose (raisins contain lots of glucose) and fructose (honey); by definition, carbohydrates have multiple −OH group, there are also polymeric carbohydrates (polysaccharides) like cellulose and starch; (3) amino acids (components of proteins) which always contain N, C, O and H; and (4) nucleotides combined from carbon cycle with nitrogen (heterocycle), sugar, and phosphoric acid; polymeric nucleotides are nucleic acids such as DNA and RNA.

On the next page is the most important thing for everyone who need to learn chemistry (and yes, you need to): periodic table. It was invented by Dmitry Mendeleev in 1869 and updated since, mostly by adding newly discovered and/or synthesized elements. Note that Roman numerals were added to standard table to show numbers of *main groups*.

1 IA

18 VIIIA

| 1 1.0079 H Hydrogen | | | | | | | | | | | | | | | | | 2 4.0025 He Helium |

2 IIA

13 IIIA 14 IVA 15 VA 16 VIA 17 VIIA

| 3 6.941 Li Lithium | 4 9.0122 Be Beryllium | | 5 10.811 B Boron | 6 12.011 C Carbon | 7 14.007 N Nitrogen | 8 15.999 O Oxygen | 9 18.998 F Flourine | 10 20.180 Ne Neon |

| 11 22.990 Na Sodium | 12 24.305 Mg Magnesium | 13 26.982 Al Aluminium | 14 28.086 Si Silicon | 15 30.974 P Phosphorus | 16 32.065 S Sulphur | 17 36.453 Cl Chlorine | 18 39.948 Ar Argon |

3 IIIA 4 IVB 5 VB 6 VIB 7 VIIB 8 VIIIB 9 VIIIB 10 VIIIB 11 IB 12 IIB

| 19 39.098 K Potassium | 20 40.078 Ca Calcium | 21 44.956 Sc Scandium | 22 47.867 Ti Titanium | 23 50.942 V Vanadium | 24 51.996 Cr Chromium | 25 54.938 Mn Manganese | 26 55.845 Fe Iron | 27 58.933 Co Cobalt | 28 58.693 Ni Nickel | 29 63.546 Cu Copper | 30 65.39 Zn Zinc | 31 69.723 Ga Gallium | 32 72.64 Ge Germanium | 33 74.922 As Arsenic | 34 78.96 Se Selenium | 35 79.904 Br Bromine | 36 83.8 Kr Krypton |

| 37 85.468 Rb Rubidium | 38 87.62 Sr Strontium | 39 88.906 Y Yttrium | 40 91.224 Zr Zirconium | 41 92.906 Nb Niobium | 42 95.94 Mo Molybdenum | 43 96 Tc Technetium | 44 101.07 Ru Ruthenium | 45 102.91 Rh Rhodium | 46 106.42 Pd Palladium | 47 107.87 Ag Silver | 48 112.41 Cd Cadmium | 49 114.82 In Indium | 50 118.71 Sn Tin | 51 121.76 Sb Antimony | 52 127.6 Te Tellurium | 53 126.9 I Iodine | 54 131.29 Xe Xenon |

| 55 132.91 Cs Caesium | 56 137.33 Ba Barium | 57-71 Lanthanide | 72 178.49 Hf Hafnium | 73 180.95 Ta Tantalum | 74 183.84 W Tungsten | 75 186.21 Re Rhenium | 76 190.23 Os Osmium | 77 192.22 Ir Iridium | 78 195.08 Pt Platinum | 79 196.97 Au Gold | 80 200.59 Hg Mercury | 81 204.38 Tl Thallium | 82 207.2 Pb Lead | 83 208.98 Bi Bismuth | 84 209 Po Polonium | 85 210 At Astatine | 86 222 Rn Radon |

| 87 223 Fr Francium | 88 226 Ra Radium | 89-103 Actinide | 104 261 Rf Rutherfordium | 105 262 Db Dubnium | 106 266 Sg Seaborgium | 107 264 Bh Bohrium | 108 277 Hs Hassium | 109 268 Mt Meitnerium | 110 281 Ds Darmstadtium | 111 280 Rg Roentgenium | 112 285 Cn Copernicium | 113 286 Nh Nihonium | 114 289 Fl Flerovium | 115 289 Mc Moscovium | 116 293 Lv Livermorium | 117 294 Ts Tennessine | 118 294 Og Oganesson |

Legend:
- Alkali Metal
- Alkaline Earth Metal
- Metal
- Metalloid
- Non-metal
- Halogen
- Noble Gas
- Lanthanide / Actinide

| Z mass Smb Name | man-made |

| 57 138.91 La Lanthanum | 58 140.12 Ce Cerium | 59 140.91 Pr Praseodymium | 60 144.24 Nd Neodymium | 61 145 Pm Promethium | 62 150.36 Sm Samarium | 63 151.96 Eu Euidpium | 64 157.25 Gd Gadolinium | 65 158.93 Tb Terbium | 66 162.50 Dy Dysprosium | 67 164.93 Ho Holmium | 68 167.26 Er Erbium | 69 168.93 Tm Thulium | 70 173.04 Yb Ytterbium | 71 174.97 Lu Lutetium |

| 89 227 Ac Actinium | 90 232.04 Th Thorium | 91 231.04 Pa Protactinium | 92 238.03 U Uranium | 93 237 Np Neptunium | 94 244 Pu Plutonium | 95 243 Am Americium | 96 247 Cm Curium | 97 247 Bk Berkelium | 98 251 Cf Californium | 99 252 Es Einsteinium | 100 257 Fm Fermium | 101 258 Md Mendelevium | 102 259 No Nobelium | 103 262 Lr Lawrencium |

Periodic table.

3.4 Very Basic Features of Life

There are two basic features of all Earth life (and we do not actually know any other Life):

1. Semi-permeable membranes:

   Most of living cells surrounded by oily cover which consists of two layers made of lipids and embedded there proteins. These membranes allow some molecules (e.g., gases or lipids) to go through, but most molecules only enter "with permission", under the tight control of membrane proteins.

2. DNA → RNA → proteins:

   That sequence is called *transcription* (first arrow) and *translation* (second arrow). DNA stores information in form of nucleotide sequence, then fragments of DNA are copying into RNA (transcription). RNA, in turn, controls protein synthesis (translation). This is sometimes called the "central dogma of molecular biology".

Consequently, to solve problem of the origin of life, we need to solve the origin of these two basic features, and also to understand how did it happen that they merged together in the system which we call *living cell*.

3.5 How to Be the Cell

Most simple, prokaryotic cell should perform several essential duties in order to survive. These are:

1. Obtaining energy. In all living world, the energy is accumulated in the form of ATP molecules. To make ATP, there are three most common ways:

   Phototrophy Energy from the light of Sun.

   Organotrophy Energy from burning of organic molecules, either slow (fermentation), or fast (respiration).

   Lithotrophy Energy from inorganic chemical reactions ("rocks").

2. Obtaining building blocks (monomers which are using to built polymers like nucleic acids, proteins and polysaccharides). The principal monomer in the living world is glucose. From glucose, it is possible to chemically create everything else (of course, one must add nitrogen and phosphorous when needed). There are two principal ways to obtain monomers:

   Autotrophy Make monomers from carbon dioxide.

71

Heterotrophy Take monomers from somebody else's organic molecules.

There are six possible combinations of these above processes. For example, what we called "photosynthesis" is in fact **photoautotrophy**. Prokaryotes are famous because they have all six combinations at work.

3. Multiply. There are always three steps:

   (a) Duplicate DNA. As it is a double spiral, one must unwind it, and then build the antisymmetric copy of each chain in accordance with a simple complement rule—each nucleotide make hydrogen bond only with one nucleotide of other type:

A	T
T	A
G	C
C	G

   (b) Split duplicated DNA.

   (c) Split the rest of the cell.

   Prokaryotic DNA is small and circular, optimized for the speedy duplication and division. Consequently, prokaryotes multiply with alarming speed.

4. Make proteins. This process involves transcription ans translation. As proteins are "working machines" of the cell and DNA is an "instruction book", there must be the way to transfer this information from DNA to proteins. It usually involves RNA which serves as temporary "blueprints" for proteins:

   (a) DNA and RNA each contains four types of nucleotides, this is an alphabet.

   (b) With help of enzymes, pieces of DNA responsible for one protein (gene) copied into RNA. Rules are almost the same as for DNA duplication above, but T from DNA is replaced in RNA with U.

   (c) The sequence of nucleotides is a language in which every tree nucleotides mean one amino acid.

   (d) Ribosomes translate trios of nucleotides (triplets) into amino acids and make proteins. They do it in accordance with *genetic code*:

	U		C		A		G		
U	UUU	Phe	UCU	Ser	UAU	Tyr	UGU	Cys	U
	UUC	Phe	UCC	Ser	UAC	Tyr	UGC	Cys	C
	UUA	Leu	UCA	Ser	UAA	*STOP*	UGA	*STOP*	A
	UUG	Leu	UCG	Ser	UAG	*STOP*	UGG	Trp	G
C	CUU	Leu	CCU	Pro	CAU	His	CGU	Arg	U
	CUC	Leu	CCC	Pro	CAC	His	CGC	Arg	C
	CUA	Leu	CCA	Pro	CAA	Gln	CGA	Arg	A
	CUG	Leu	CCG	Pro	CAG	Gln	CGG	Arg	G
A	AUU	Ile	ACU	Thr	AAU	Asn	AGU	Ser	U
	AUC	Ile	ACC	Thr	AAC	Asn	AGC	Ser	C
	AUA	Ile	ACA	Thr	AAA	Lys	AGA	Arg	A
	AUG	Met	ACG	Thr	AAG	Lys	AGG	Arg	G
G	GUU	Val	GCU	Ala	GAU	Asp	GGU	Gly	U
	GUC	Val	GCC	Ala	GAC	Asp	GGC	Gly	C
	GUA	Val	GCA	Ala	GAA	Glu	GGA	Gly	A
	GUG	Val	GCG	Ala	GAG	Glu	GGG	Gly	G

Genetic code. All amino acids designated with shortcuts.

5. Make sex. To evolve, organisms must diversify first, natural selection works only if there is an initial diversity. There are two ways to diversify:

   (a) *Mutations* which are simply mistakes in DNA. Majority of mutations are bad, and many are lethal. The probability to obtain useful mutation is comparable with probability to mend your cell phone using hammer.

   (b) *Recombinations* are much safer, they increase diversity but unable to create novelties. In addition, recombinations serve also as a way to discard bad genes from the "gene pool" of population because from time to time, two or more bad genes meet together in one genotype and this combination becomes lethal.

Prokaryotes developed bacterial conjugation when two cell exchange parts of their DNA, this facilitates recombination. In bacterial world, recombination is possible not only within one population, but sometimes also between different species, this is called *horizontal gene transfer*.

3.6   Overview of the Cell

This small section will schematically describe structures of two cells, cell of eukaryote, and cell of prokaryote.

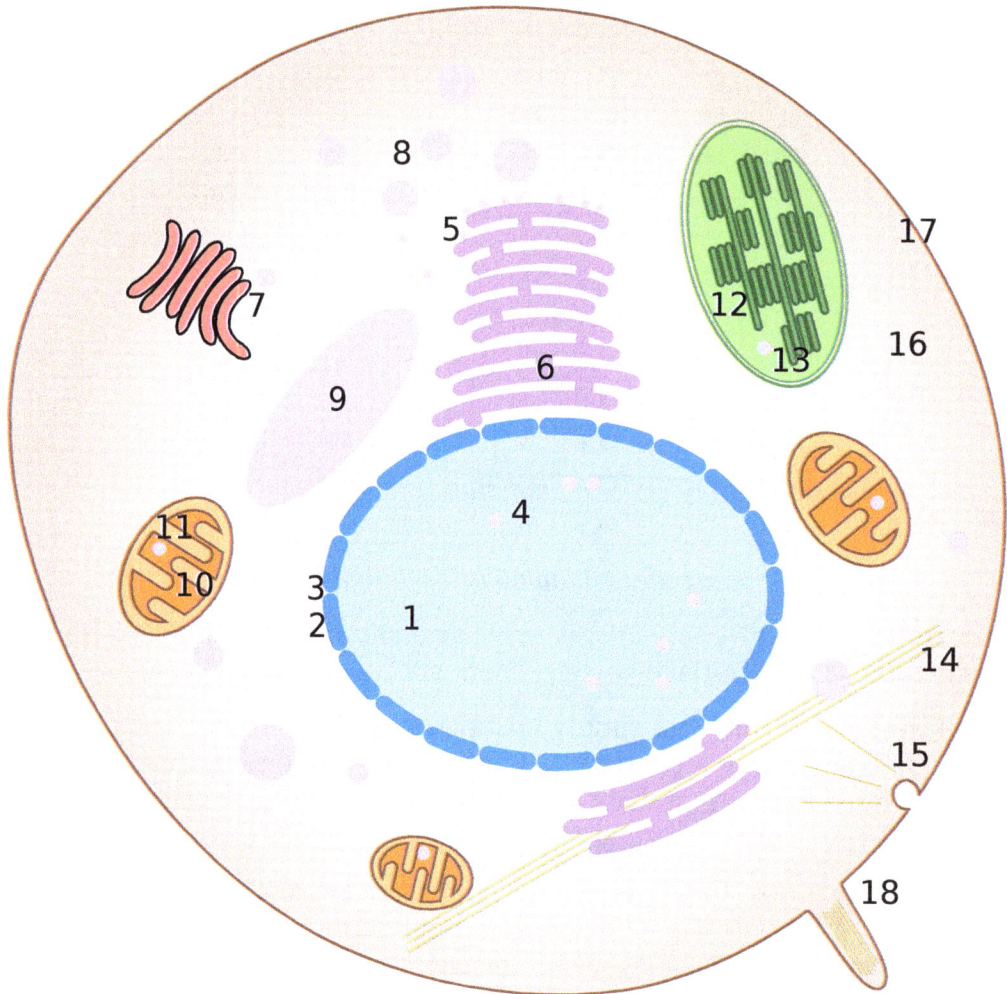

Cell of eukaryote, second-order cell: 1 nucleus, 2 nuclear envelope, 3 nuclear pore, 4 DNA = chromatin = chromosomes, 5 ribosomes (black), 6 ER, 7 AG, 8 vesicles, 9 vacuole (big vesicle), 10 mitochondria (surrounded with double membrane), 11 mitochondrial DNA, 12 chloroplast (surrounded with double membrane), 13 chloroplast DNA, 14 cytoskeleton (brown), 15 phagocytosis (caught in the middle of process), 16 cytoplasm, 17 cell membrane, 18 eukaryotic flagella.

Prokaryotic cell, cell of Monera, is much smaller, much more rigid and much simpler. Labels not provided because there is not much to label, and what is available, was already shown in the eukaryote. Except the *cell wall* which is outside of the cell membrane (*some* eukaryotes have cell wall though), and prokaryotic flagella (right bottom corner) which is just a molecule of protein:

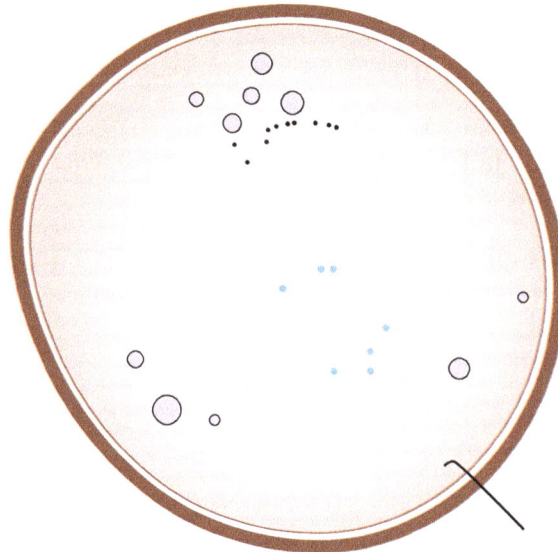

Yes, this is all.

3.7 Ecological Interactions: Two-Species Model

Two-species model allows to describe how two theoretical species might influence each other. For example, Species I may facilitate Species II: it means that if biomass (sum of weight) of Species I increases, biomass of Species II also increases (+ interaction). There are also + and 0 interactions. Two species and three signs make **six** combinations:

	+	0	−
+	mutualism	commensalism[1]	exploitation[2]
0	...	neutralism	amensalism
−	...	...	interference[3]

[1] Includes phoresy (transportation), inquilinism (housing) and "sponging".
[2] Includes predation, parasitism and phytophagy.
[3] Includes competition, allelopathy and aggression.

Mutualism It sometimes called "symbiosis". Two different species collaborate to make each other life better. One of the most striking example is lichenes which is algae-fungus mutualism.

Commensalism Remember "Finding Nemo"? Clown fish lives inside actinia. This type of commensalism is called "*housing*". Another example is suckerfish and shark, this is *phoresy*. *Sponging* happens when scavengers feed on what is left after the bigger carnivore meal.

Exploitation This is the most severe interaction. *Predation* kills, but *parasitism* or *phytophagy* (the only difference is that second uses plants) do not.

Neutralism Rare. Philosophically, everything is connected in nature, and if Species I and II live together, they usually interact, somehow.

Amensalism This happens when suppressing organism is, for example, much bigger then the "partner". Big trees often suppress all surrounding smaller plants.

Interference *Competition* happens when Species I and II share same ecological niche, have similar requirements. *Gause's Principle* says that sooner or later, one of them wins and another looses. *Allelopathy* is a mediated competition, typically through some chemicals like antibiotics. Most advanced (but least pleasant) is the direct *aggression* when individuals of one species physically eliminate the other one.

3.8   How to become an animal

Three driving forces of eukaryotic evolution:

- Prey and predator interactions
- Surface and volume
- Ecological pyramid (Fig. 3.1).

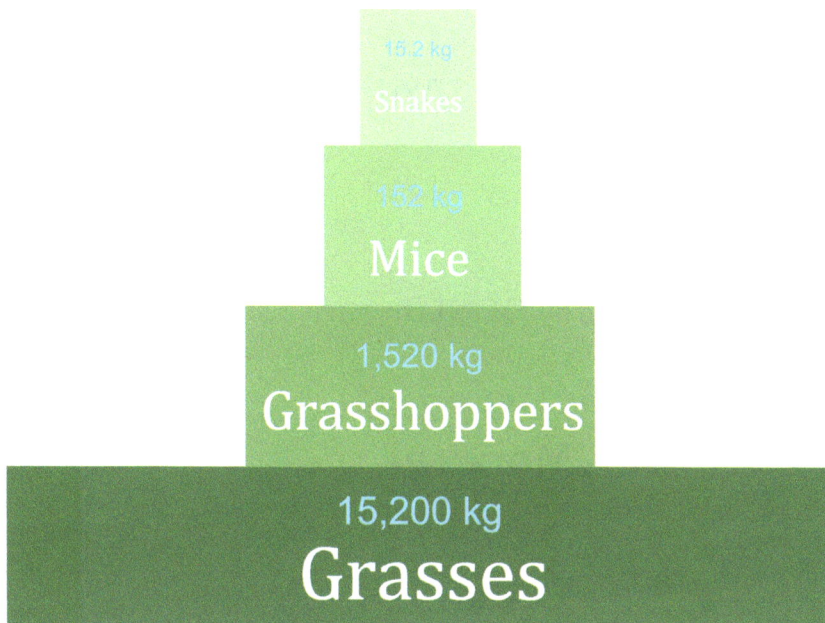

Figure 3.1: Ecological pyramid, or biomass pyramid

finally resulted in appearance of animals and plants. But these two groups originated differently.

Animals belong to the highest level of the pyramid of life (Fig. 3.2). They not only multi-cellular but also multi-tissued creatures. It is easy to become multicellular, enough is not to split cells completely after mitosis. And the big advantage is immediately feel: size.

To make a big body, it is much easier to join several cells then grow one cell. Explanation lays (as well as explanation of many other biology phenomena) in the *surface/volume paradox*: the more is the volume, the less is relative surface, and this is often bad. Multiplication of cells allows to be big without decreasing relative surface. And being big is a good idea for many living things, especially for plants (the more is the size, the more intensive is photosynthesis) and for prey in general (the bigger is prey, the more chances to survive after contact with predator).

But this is not working out of the box for *active hunters* like animals' ancestors! To move, they need also the tight coordination between cells, and to eat, they need altruistic cells which feed other cells. Consequently, first animals (like *phagocytella*: Fig. 3.3) must acquire have at least two tissues: (1) surface cells, adapted to motion, likely flagellate, and (2) cells located in deeper layers, adapted to digestion, probably amoeba-like. This evolution could go through several stages:

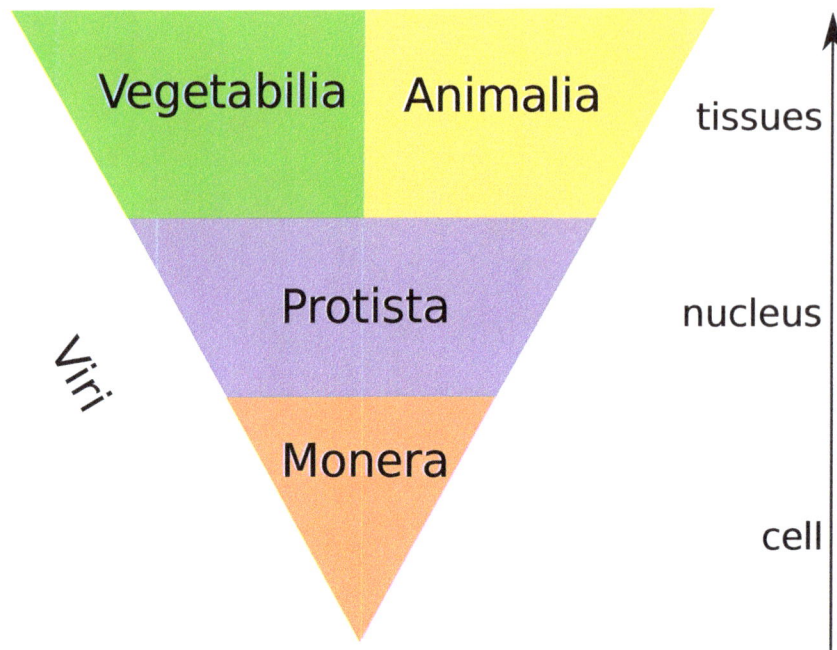

Figure 3.2: "Pyramid of Life."

- Blastaea: not the animal yet. *Volvox*, *Proterospongia*.

- Phagocytella. Two tissues: *kinoblast* and *phagocytoblast*. *Trichoplax*.

- Gastraea (jellyfish without "bells and whistles"). Three tissues: ectoderm, entoderm and mesoderm. Closed gut.

What about communication, circulation, gas exchange etc.? If the first animal was small enough, all of these will run without specialized tissues, via diffusion, cell contacts and so on. But when size grows, the surface/volume paradox dictates that new and new tissue and organ appear. And size surely will grow because there is a constant race of arms between prey and predator, and between different predators.

3.9 How to be an animal

More organized, bigger animal has multiple needs, and therefore, multiple tissues and organs (Fig. 3.4):

- locomotion: appendages, skin-muscular bag (A), fins;

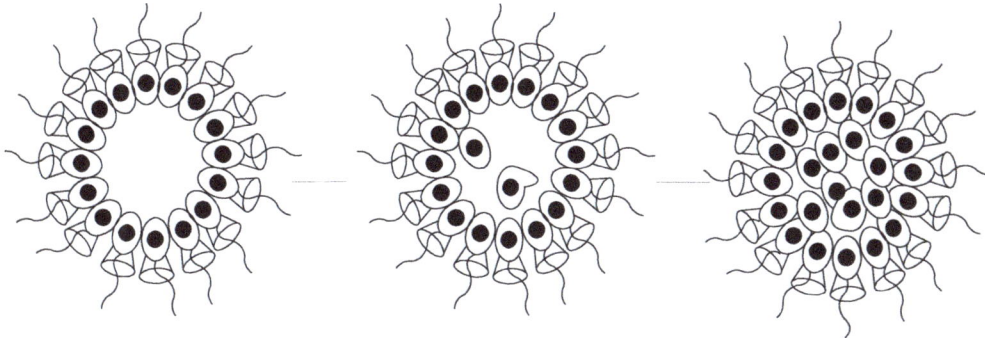

Figure 3.3: From blastaea to phagocytella: hypothetical scenario. See explanations of these terms in the text.

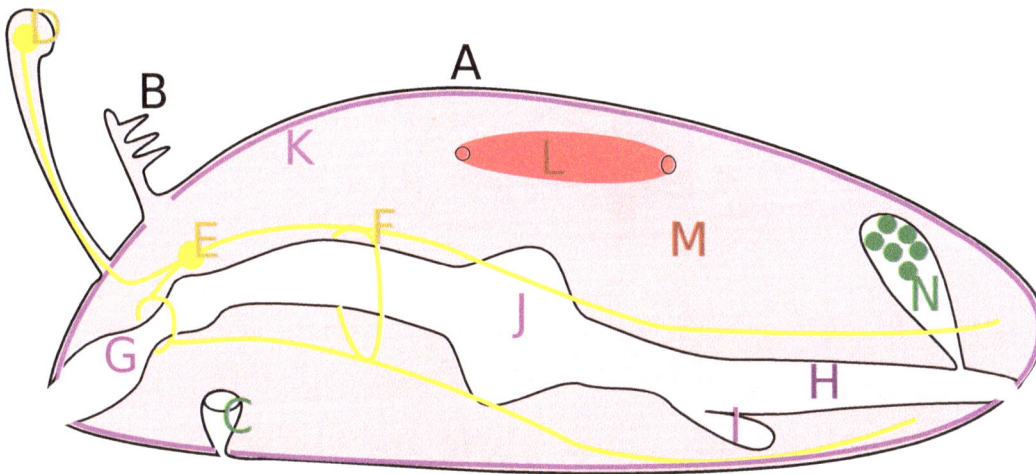

Figure 3.4: "Ideal worm." See the text for explanation of letters.

- support: many types of skeleton (endoskeleton, chitinous exoskeleton, shells, skin plates) and hydrostatic skeleton based on body cavities filled with liquid (A + K + M);

- feeding and excretion: mouth, anus, intestines, pharynx (G), stomach (J), digestion glands (like liver, I) *etc.*;

- osmoregulation: simple nephridia (C) and complicated kidneys;

- gas exchange: external gills (B), internal lungs and tracheas;

- circulation: open (M) or closed blood system with hear(s) (L);

- reception: eyes (D), mechanical sensors (ears, hairs), chemical sensors (nose) and many others;

- communication: neural cells (neurons), nerves (groups of neurons) (F), ganglia (E) and brain (masses of neurons);

- reproduction: sexual organs filled with sexual cells (N), male and female, separately or together, and fertilization "tools".

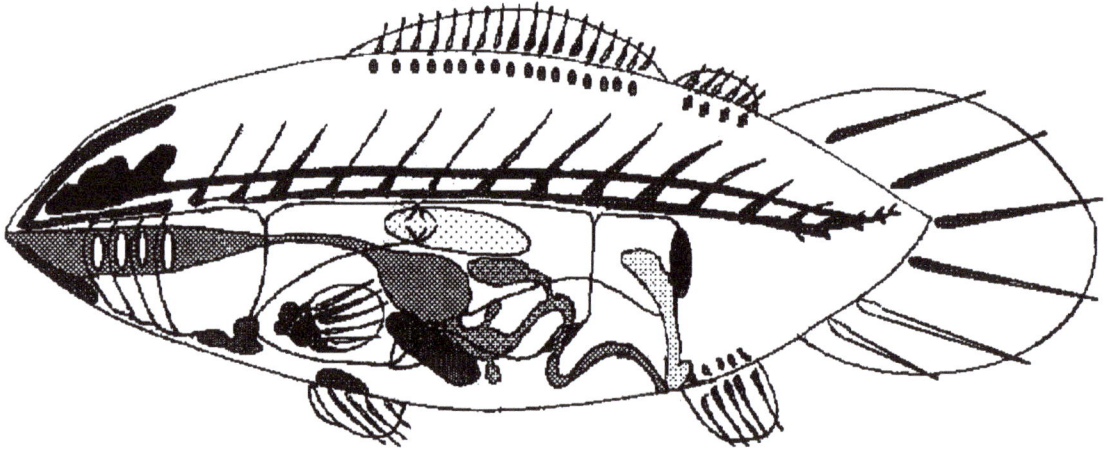

Figure 3.5: Body of the vertebrate: compare with the worm above.

3.10 Animalia: body plans, phyla and classes

From Georges Cuvier times, highest animal groups (phyla) are understood as different *body plans* (Fig. 3.7).

There are some important animal phyla with notes about their body plans. Note that most of mentioned characters do not belong to the 100% of phylum species. As biology is a science of exceptions, this is normal. The following table lists many of animal phyla and also classes of chordates. Look also on "split pyramid" scheme (Fig. 3.6).

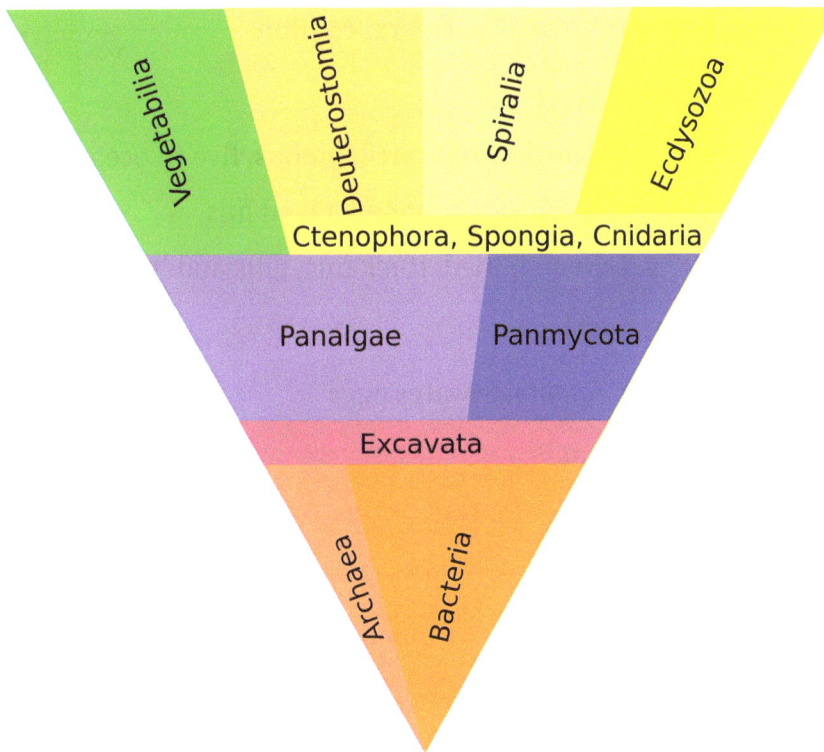

Figure 3.6: "Pyramid of Life", with more splits

Subregnum *Spongia:* no symmetry

> Phylum 1. Porifera: sitting filtrators with skeletal spicules (H)

Subregnum *Cnidaria:* radial symmetry, stinging cells

> Phylum 2. Anthozoa: sitting, colonial, with skeleton (I)

> Phylum 3. Medusozoa: swimming, solitary, soft (A)

Subregnum *Bilateria:* bilateral symmetry, likely originated from crawling habit

Infraregnum Deuterostomia: with specific embryogenesis

> Phylum 4. Echinodermata: small-plate exoskeleton, secondary radial, water-vascular appendages (E)

> Phylum 5. Chordata: head and tail, gills in pharynx, axial skeleton (B)

Subphylum *Cephalochordata:* Acrania (no skull)

> Classis 1. *Leptocardii:* lancelet with no eyes and jaws

Subphylum *Vertebrata:* vertebral column

Pisces: fish-like, gills

Classis 2. *Chondrichtyes:* cartilaginous, live in ocean

3. *Actinopterygii:* boned, rayed fins

4. *Dipnoi:* boned, thick fins, gills and lungs

Tetrapoda: four legs

Superclassis *Anamnia:* water eggs

Classis 5. *Amphibia*

Superclassis *Amniota:* terrestrial eggs

Classis 6. *Reptilia:* no feathers, four-pedal

7. *Aves:* feathers, bipedal

8. *Mammalia:* grinding jaws, fur, milk

Infraregnum Protostomia: specific embryogenesis

Superphylum *Spiralia:* worms, mollusks and alike

Phylum 6. Mollusca: shell, body straight (F)

Phylum 7. Lophophorata: shell and alike, body curved (J)

Phylum 8. Annelida: naked, segmented (G)

Superphylum *Ecdysozoa:* molting, with chitinous cuticle

Phylum 9. Nemathelminthes: worms with bending motion and primary cavity (C)

Phylum 10. Arthropoda: both body and appendages segmented (D)

Classis 1. *Chelicerata:* spiders, ticks, mites, scorpions

2. *Malacostraca:* crabs, lobsters, shrimp

3. *Hexapoda:* insects and alike

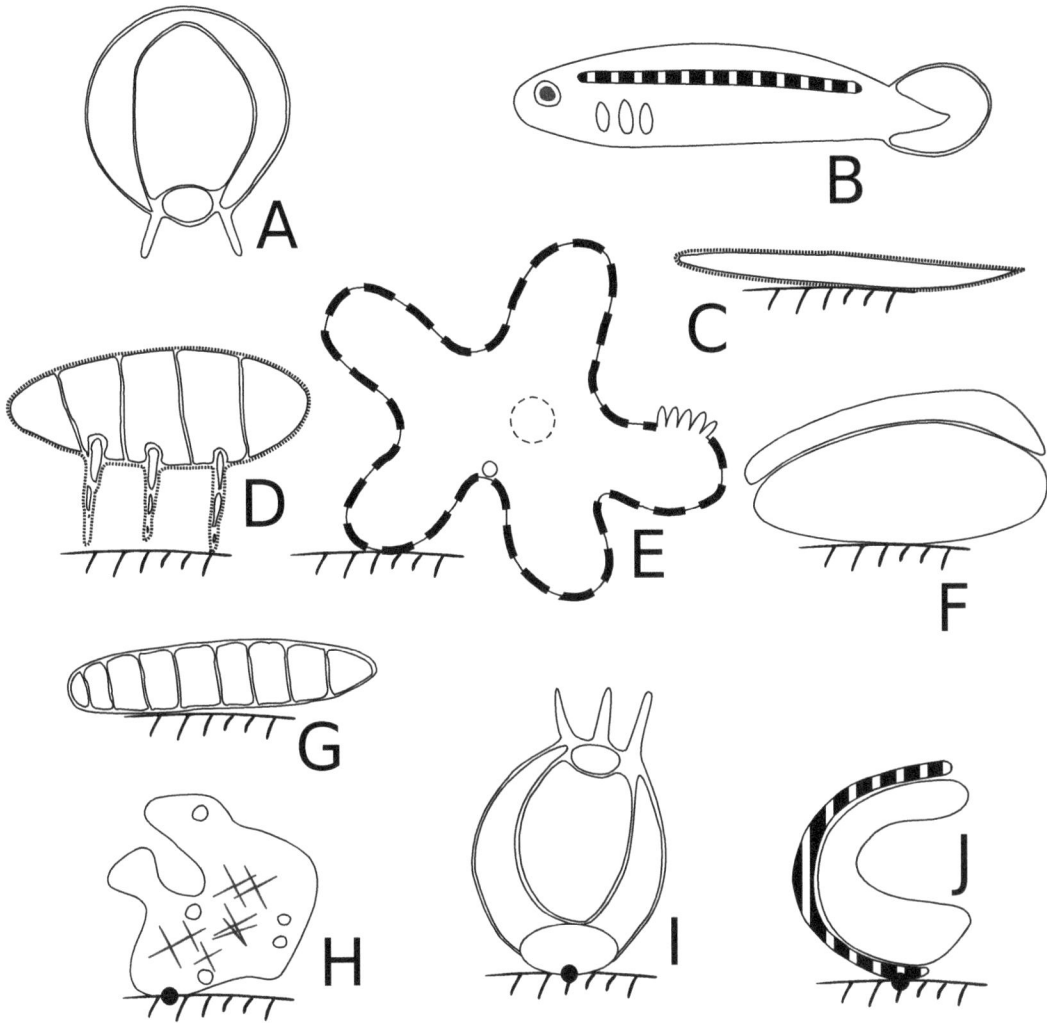

Figure 3.7: Ten animal body plans, corresponding with phyla. See the text for explanation of letters.

3.11 Peas and flies: basics of inheritance

3.11.1 Syngamy and meiosis

- Syngamy if one of safest mechanisms of recombination

- Continuous syngamy is not good

- Meiosis is the counterpart of syngamy

- The simplest way to make meiosis is to split paired (homologous) chromosomes between two daughter cells, and then *immediately* (without S-phase) split duplicated DNA; thus meiosis has two divisions

3.11.2 Male and female

In syngamy, there are always two partners. They could be superficially almost identical—only genotypes and some surface chemistry differ. Two dangers are here: to mate with same genotype (then why to mate?) or mate with other species (which could result in broken genotype). To improve partner recognition, there are many mechanisms.

One is based simply on size. If one is bigger, then smaller one is presumably the good partner. From this point, smaller cell is called male, and bigger called female.

Now, females could invest in storage (and bigger size) whereas males invest in numbers. This strategy will dramatically improve fertilization and also allows to select better males. It results in big, non-motile female cells and small, fast-moving, numerous male cells. Here females called oocytes (or egg cells) and males—spermatozoa.

The ultimate step could be non-motile males too, but this is not frequent because they will need the external help for the fertilization. These non-motile male cells (spermatia) exist in red algae, sponges, crustaceans and flowering plants.

3.11.3 Mendel's theory, explanations and corrections

- In crossings, he often used *two* different variants of one character:
  *two **genotype** variants (**alleles**, paralogs) of one **gene** which control two variants of one **phenotype***

- "Factors" (genes) are paired in plant but **separated in gametes**:
  because of meiosis

- One "factor" is **dominant**:
  one variant is working DNA, the other is not

- Different characters are separating between gametes **independently**:
  this is how anaphase I of meiosis goes

- This is because different characters are located in **different places**:
  i.e., in different pairs of chromosomes

- If genes are located in the same chromosome, they are **linked** and will not be inherited independently

- However, linkage could be broken in **crossing-over** (it runs in prophase I of meiosis)

- Sometimes, sex is determined with chromosome set: one gender has the pair where chromosomes are non-equal

3.11.4   Anaphase I and recombinants

Imagine that parent is fully heterozygous, like in Mendel's first generation. It has red flowers (Rr) and long stems (Ll), the whole genotype is then "RrLl".

There are **two possibilities** in the anaphase I:

1. Either "R-chromosome" and "L-chromosome" come together to one pole (consequently, "l-chromosome" and "r-chromosome" to other pole)

2. Or "R-chromosome" + "l-chromosome" come one way, and "r-chromosome" + "L-chromosome" another way

Each variant has 1/2 (50%) probability, like in throwing a coin.

Four gamete types are possible:

1. RL
2. rl
3. Rl
4. rL

Four gametes give 16 combinations (phenotypes are short-stemmed, **long-stemmed**, white-flowered and red-flowered):

	RL	rl	Rl	rL
RL	**RRLL**	**RrLl**	**RRLl**	**RrLL**
rl	**RrLl**	rrll	Rrll	**rrLl**
Rl	**RRLl**	Rrll	RRll	**RrLl**
rL	**RrLL**	**rrLl**	**RrLl**	**rrLL**

(Please count proportions to see that it is really 9:3:3:1.)

As R and L are dominant, only four phenotypes appear, and two of them are **recombinants**, phenotypes unlike parents.

3.12 Plant stories

3.12.1 Plants and plants

- The following factors "pushed" plants on land:
    1. Availability of light
    2. Temperature-gases conflict
    3. Increased competition in shallow waters
- Two first tissues:
    1. isolating/ventilating compound epidermis
    2. photosynthetic/storage ground tissue were response to desiccation.
- Epidermis could be developed in advance as adaptation to spore delivery.
- Next stages:
    3. supportive tissues to solve "Manhattan problem"
    4. vascular tissues to transport water and sugars
    5. branching
    5. absorption tissues (or mycorrhiza) for water uptake

3.12.2 Life cycles

This is a short list of terms associated with life cycles:

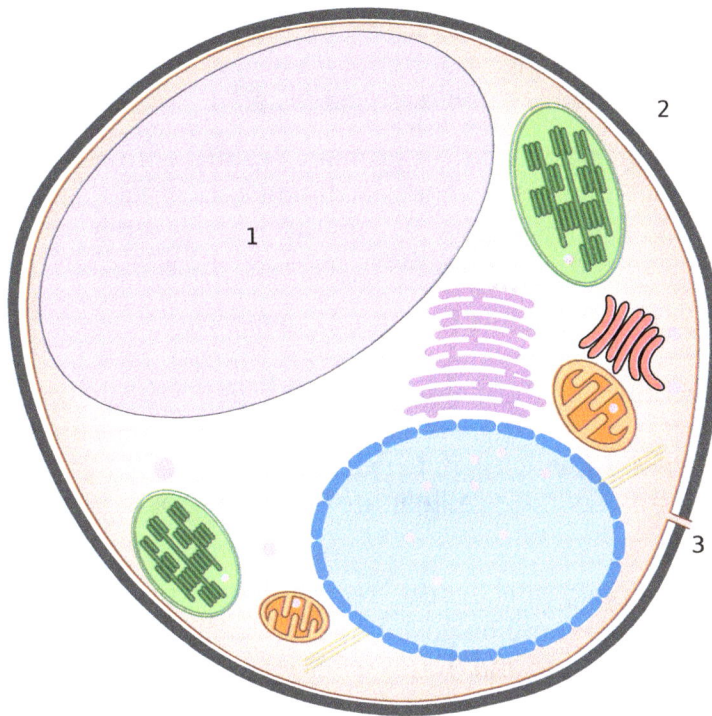

Figure 3.8: **Plant cell** (compare with two cell schemes from above): 1 vacuole, 2 cell wall, 3 plasmodesm.

- mitosis, meiosis (R!), syngamy (Y!)

- vegetative reproduction (cloning), sexual reproduction and asexual reproduction

- result of syngamy: zygote; participant of syngamy: gamete

- smaller gamete: male, bigger gamete: female; movable male gamete: spermatozoon (sperm), motionless female gamete: oocyte (egg cell)

- result of meiosis: spores

- haplont (plants: gametophyte) and diplont (plants: sporophyte)

- sporic life cycle (like in plants), gametic life cycle (like in animals) and sporic (only protists)

- sporic: gametophyte dominance (mosses) and sporophyte dominance (ferns and seed plants)

Note that Mendel "saw" genes mixed, segregated and then immediately mixed/recombined again, whereas in the life cycle of unicellular eukaryote, they are segregated, then mixed/recombined and immediately segregated again.

This is because for multicellular organism, diploid condition is better. Since not all genes are strictly dominant, then (1) diploids are broader adapted, due to two variants of gene; (2) if one copy breaks (mutation), the other still works. Diploid condition is also a handy tool to effectively segregate homozygous lethal mutations.

Syngamy is a cheapest and safest way to mix genes within population. After syngamy, the most natural step for unicellular organism is to return DNA amount back to normal, reduce it through the meiosis (of course, genes do not unmix). However, if the organism is multicellular, there is a choice because (a) they already have the developmental program allowing them to exist as stable group of cells and (b) diploid is better. So, while some of multicellular life head to meiosis, zygote of many others proceeds to diplont.

Diplont is a body of diploid cells. It still "keeps in mind" that at some point, meiosis will be required, but this could be postponed for now. Main goal for the diplont is to grow its multicellular body and (if this is reasonable), clone itself with vegetative reproduction.

Then, when time came, meiosis occurs and resulted in 4 cells. They are haploid. Here is the second choice. These new cells could proceed back to syngamy, like in animals and some protists; but in other protists and plants, these cells (now they are called spores) will grow into haplont.

Haplont is a body of haploid cells. Again, it "remembers" that at some point, syngamy will be required, but at the moment, it enjoys multicellular life which could be superficially very similar to diplont.

Finally, some cells of haplont become gametes which go to sexual reproduction, syngamy. Life cycle is now completed.

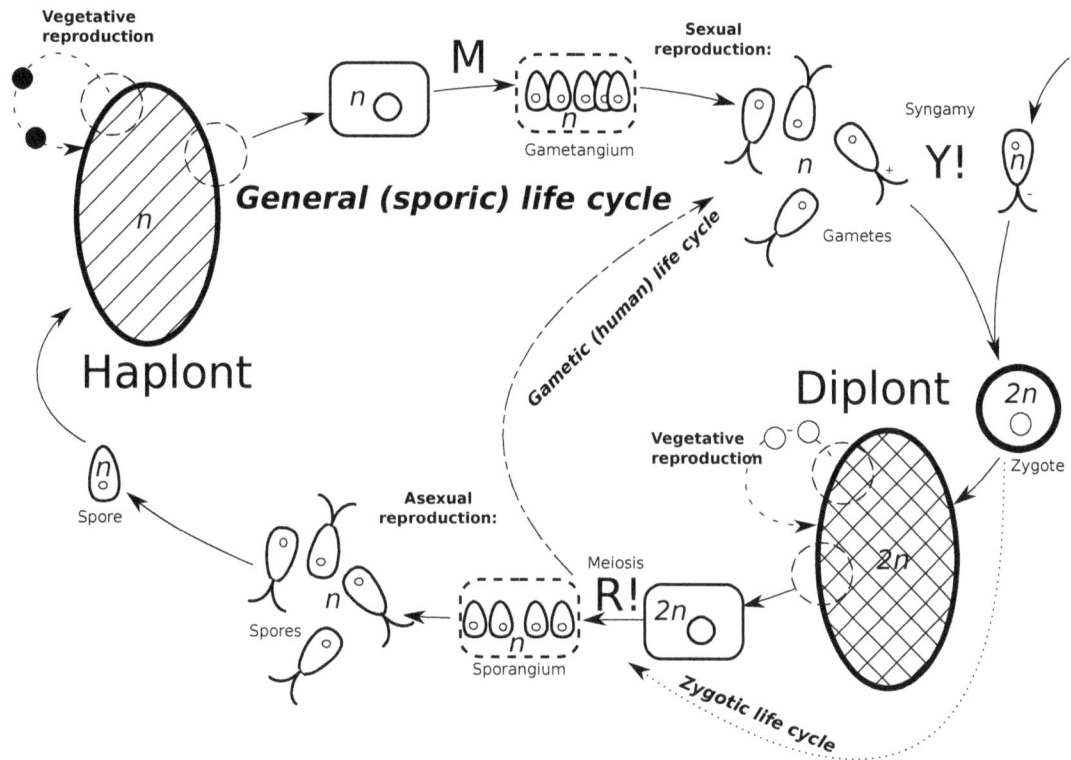

Figure 3.9: Life cycle of multicellular eukaryote.

3.12.3 Three phyla of plants

As animal phyla differ by body plans, plant phyla differ with life cycles. Plants started from green algal ancestors which have no diplont, only zygote was diploid in their life cycle. But diplont is better! So plants gradually increased the diploid stage (plant's diplont is called sporophyte) and reduced haploid (gametophyte). And they still did not reach the animal (gametic) shortcut of the life cycle, even in most advanced plants there is small gametophyte of few cells. One of reasons is that plants do not move, and young sporophyte always starts its life on the mother gametophyte.

Mosses have sporophyte which is adapted only for spore dispersal. Gametophyte is then a main photosynthetic stage which makes most of photosynthesis and therefore need to be big. However, it cannot grow big! This is because for the fertilization, it needs water. Therefore, mosses could not be larger then the maximal level of water.

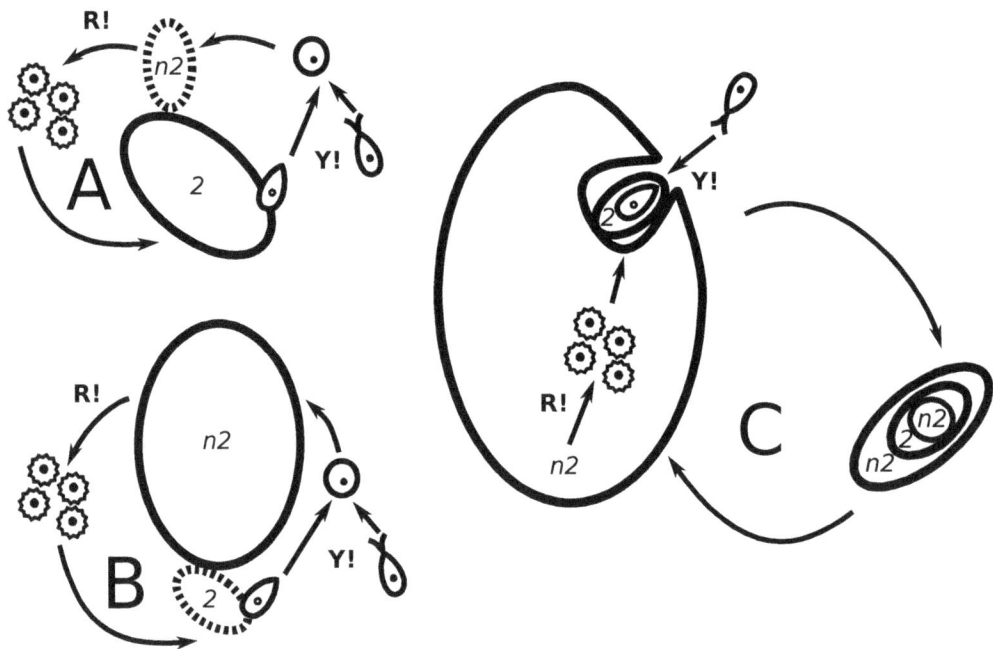

Figure 3.10: Life cycles of Bryophyta (A), Pteridophyta (B) and Spermatophyta (C).

To overcome the restriction, stages role must reverse. Ferns did that, and their gametophyte is really small and adapted only for fertilization. This works pretty well but only if the plant body is relatively small.

Trees capable to secondary growth (that is, thickening of stem with special "stem cells") will experience the ecological conflict between ephemeral, minuscule gametophyte and giant stable sporophyte. Whatever efforts sporophyte employs, result is unpredictable. Birth control, so needed for large organisms, is impossible. One solution is not to grow so big.

Another solution is much more complicated. They need to reduce gametophyte even more and place it on sporophyte. And also invent the new way of bringing males to female because between tree crowns, the old-fashioned water fertilization is obviously not possible. To make this new way (it called pollination), some other external agents must be employed. First was a wind, and the second came out of the clever trick to convert enemies into friends: insects.

Still, result was really cumbersome and the whole life cycle became much slower than in ferns, it could span years! The only way was to optimize and optimize it, until in flowering plants, it starts to be comparable and even faster then in two other phyla.

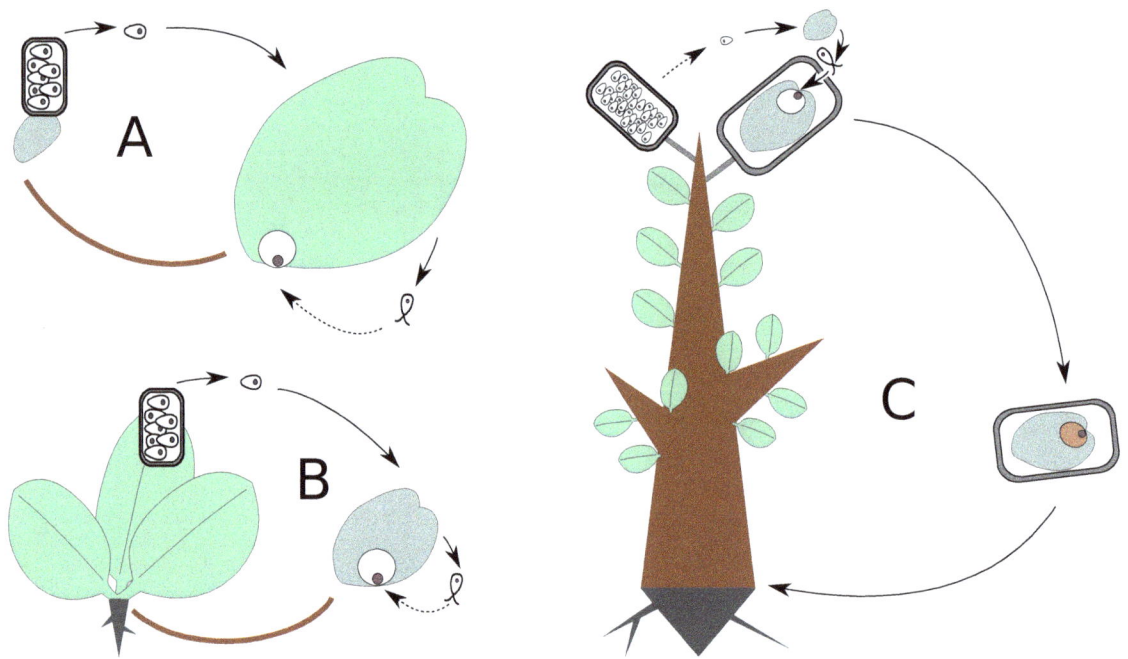

Figure 3.11: Life cycles of Bryophyta (A), Pteridophyta (B) and Spermatophyta (C): another view.

Chapter 4

Geography of Life

4.1 Ideal continent and real continents

On the next page, see the map of "ideal continent" representing the Earth land-masses, ocean currents, climates and ecoregions (according to Rjabchikov, 1960).

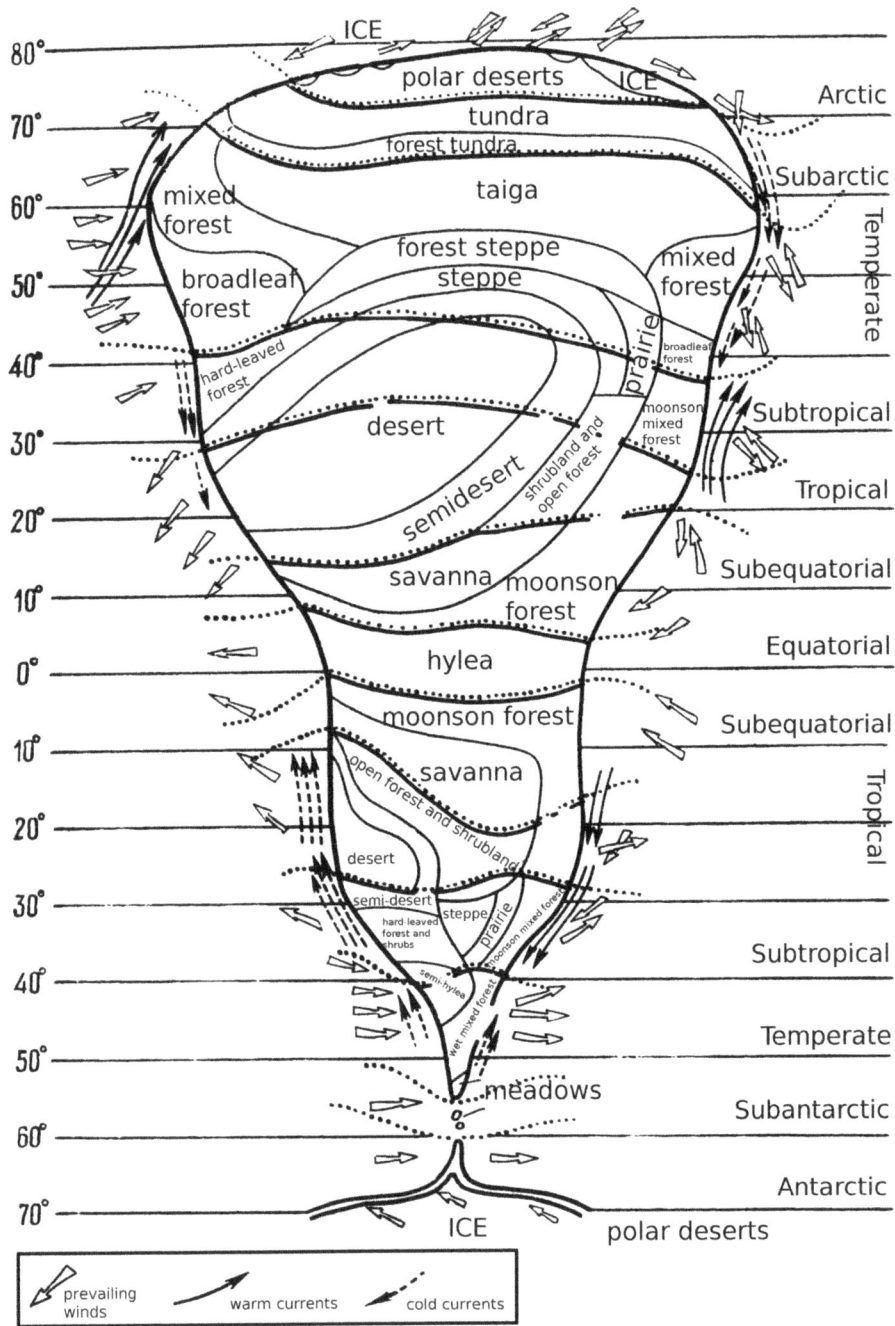

Legend:
- prevailing winds
- warm currents
- cold currents

Latitude markings (left side, top to bottom): 80°, 70°, 60°, 50°, 40°, 30°, 20°, 10°, 0°, 10°, 20°, 30°, 40°, 50°, 60°, 70°

Climate zones (right side, top to bottom): Arctic, Subarctic, Temperate, Subtropical, Tropical, Subequatorial, Equatorial, Subequatorial, Tropical, Subtropical, Temperate, Subantarctic, Antarctic

Biome labels: ICE, polar deserts, tundra, forest tundra, taiga, mixed forest, forest steppe, steppe, broadleaf forest, hard-leaved forest, desert, semidesert, shrubland and open forest, savanna, moonson forest, prairie, mixed forest, broadleaf forest, moonson mixed forest, hylea, moonson forest, savanna, open forest and shrubland, desert, semi-desert, steppe, hard-leaved forest and shrubs, prairie, moonson mixed forest, semi-hylea, wet mixed forest, meadows, ICE, polar deserts

Next few pages are filled with schematic maps of four real continents. However, these maps are simplified to the extreme in order to show the most important ecological and geographical features of the each continent.

94

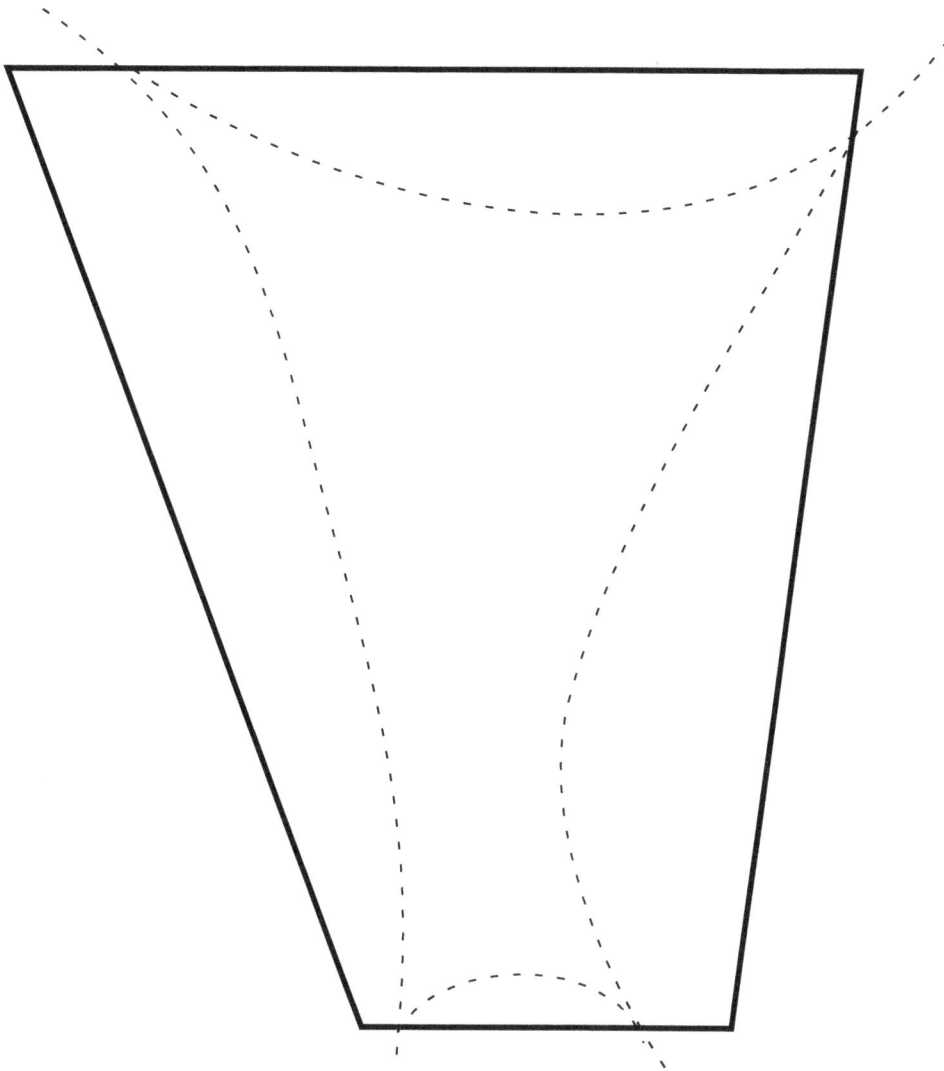

The most simple biogeographic map of North America:
cold North, wet East, hot South and dry West.

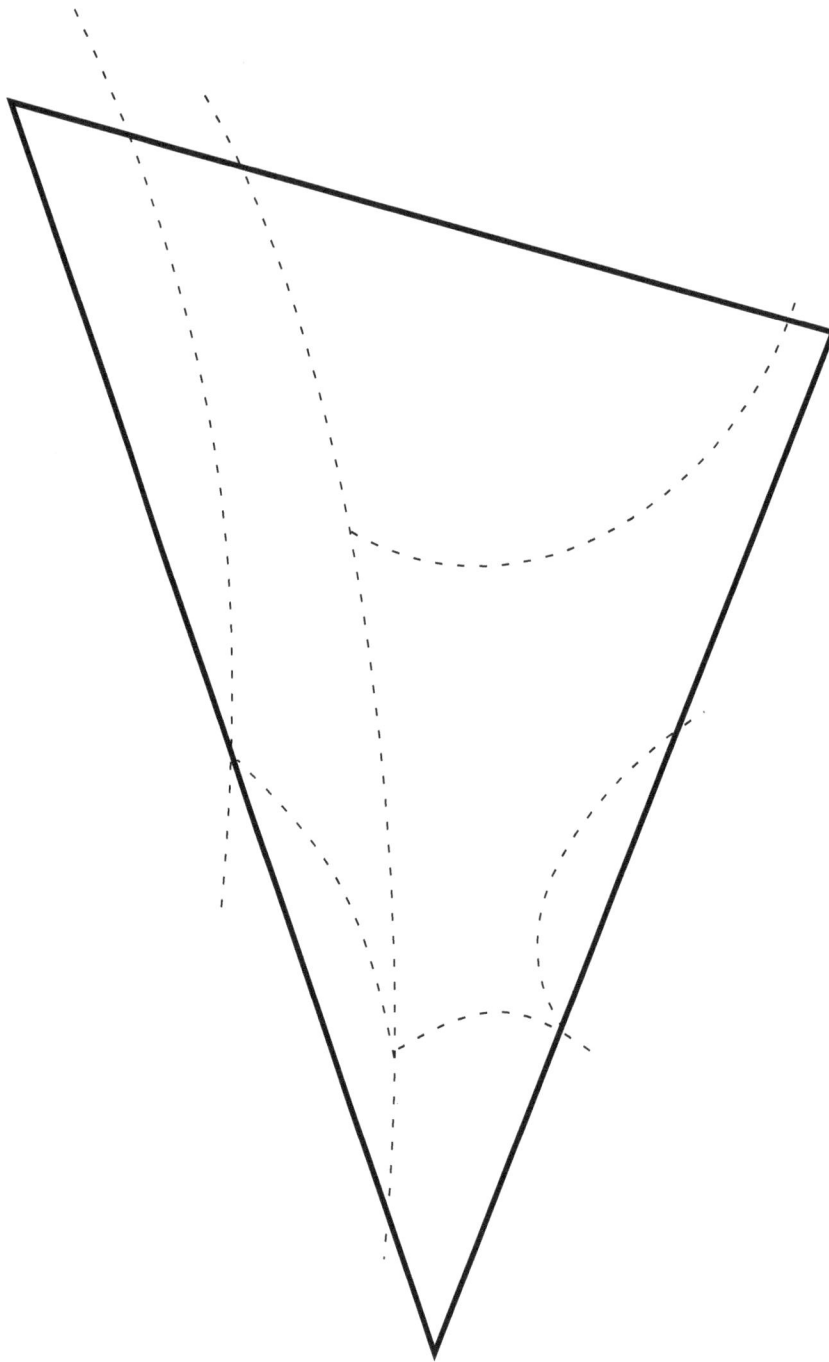

The most simple biogeographic map of South America.

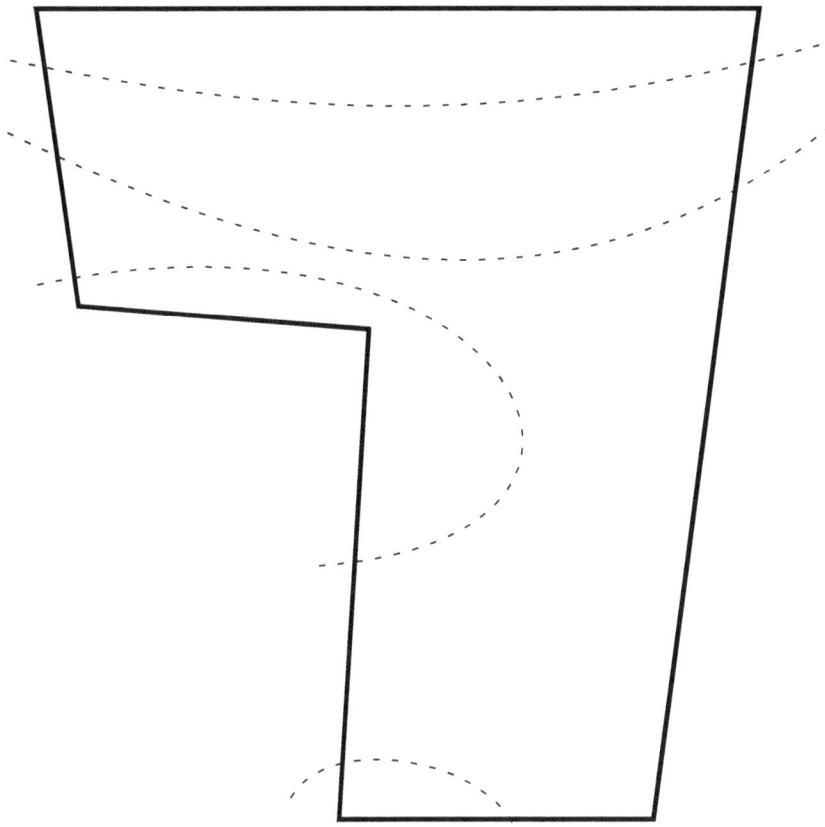

The most simple biogeographic map of Africa.

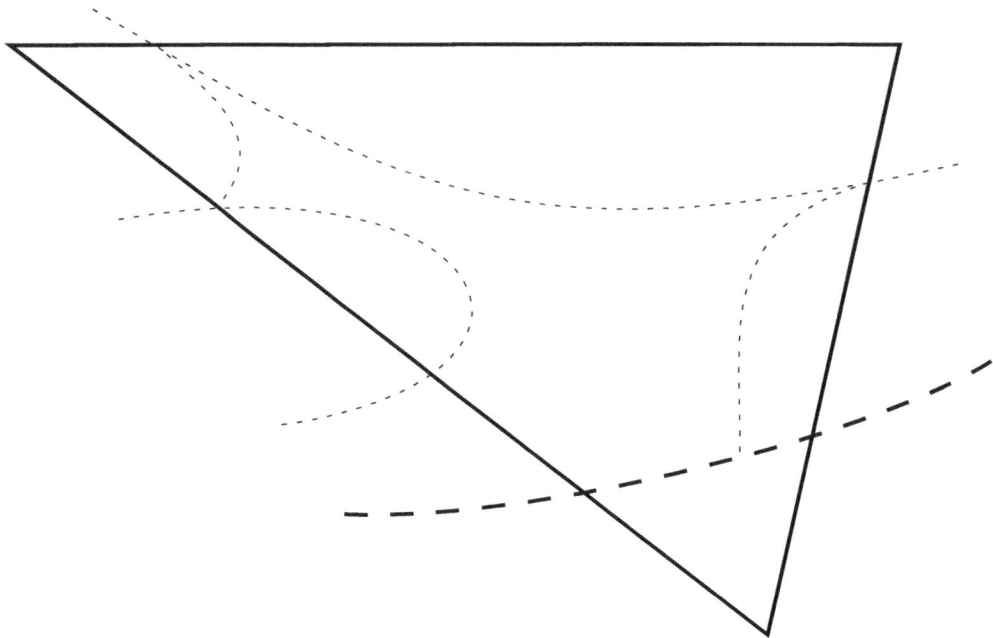

The most simple biogeographic map of Eurasia.

4.2 Architectural Models of Tropical Trees: Illustrated Key

Tropical landscape is full of trees. They rarely flower or bear fruits, and often have very similar leaves. However, shapes and structures of trunks and crowns (so similar in temperate regions) are seriously different in tropics. If you want to know tropics better, you should learn these architectural models.

The following key is based on Halle, Oldeman and Thomlinson (1978) "Tropical Trees and Forests" (pp.84–97).

1. Stem strictly unbranched (Monoaxial trees) ....2.

   – Stems branched, sometimes apparently unbranched in Chamberlain's model (polyaxial trees) ....3.

2. Inflorescence terminal ....**Holttum's model**.

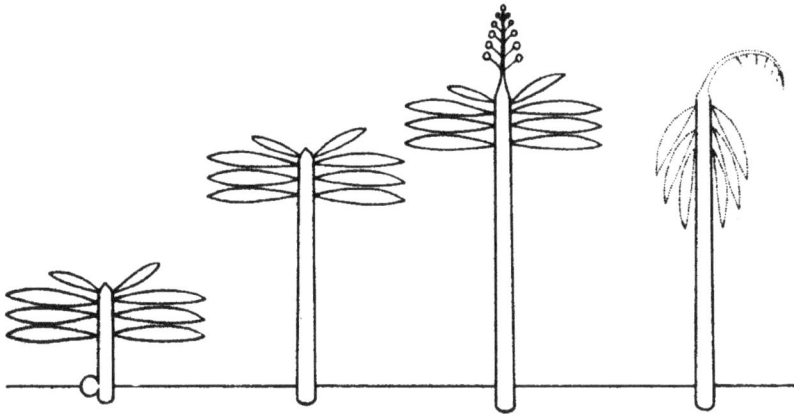

Monocotyledon: *Corypha umbraculifera* (Talipot palm—Palmae). Dicotyledon: *Sohnreyia excelsa* (Rutaceae).

– Inflorescences lateral .... **Corner's model**.

(a) Growth continuous:

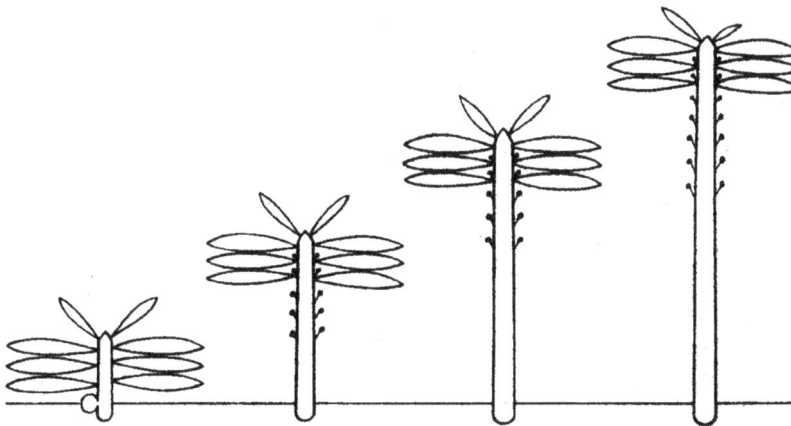

Monocotyledon: *Cocos nucifera* (coconut palm—Palmae), *Elaeis guineensis* (African oil palm—Palmae). Dicotyledon: *Carica papaya* (papaya—Caricaceae).

(b) Growth rhythmic:

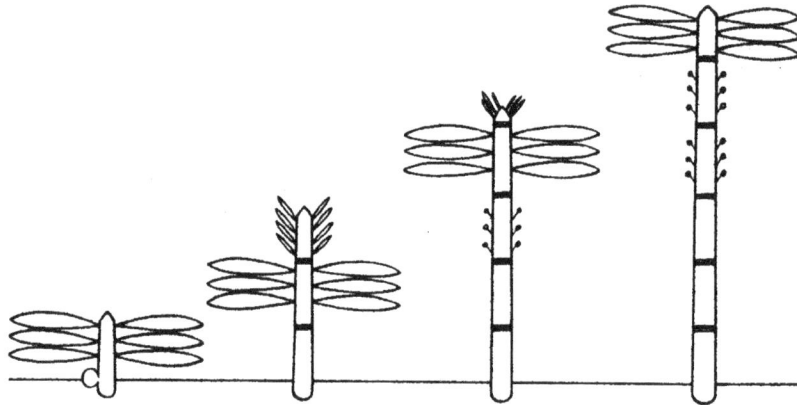

Gymnosperm: Female *Cycas circinalis* (Cycadaceae). Dicotyledon: *Trichoscypha ferntginea* (Anacardiaceae).

3 (1). Vegetative axes all equivalent, homogenous (not partly trunk, partly branch), most often orthotropic and modular . . . . 4.

— Vegetative axes not equivalent (homogenous, heterogenous or mixed but always clear difference between trunk and branches) . . . . 7.

4. Basitony, i.e., branches at the base of the module, commonly subterranean, growth usually continuous, axes either hapaxanthic or pleonanthic . . . .
. . . . **Tomlinson's model**.

(a) Hapaxanthy, i.e., each module determinate, terminating in an inflorescence:

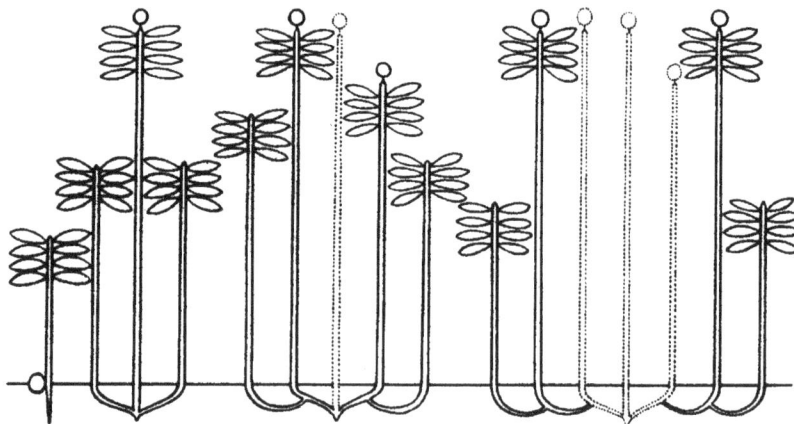

100

Monocotyledon: *Musa* cv. *sapientum* (banana—Musaceae). Dicotyledon: *Lobelia gibberoa* (Lobeliaceae).

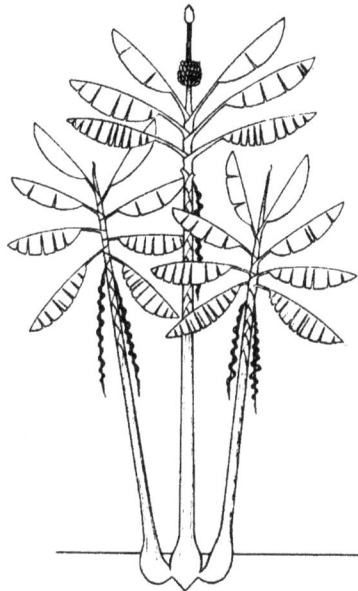

(b) Pleonanthy, i.e., each module not determinate, with lateral inflorescences

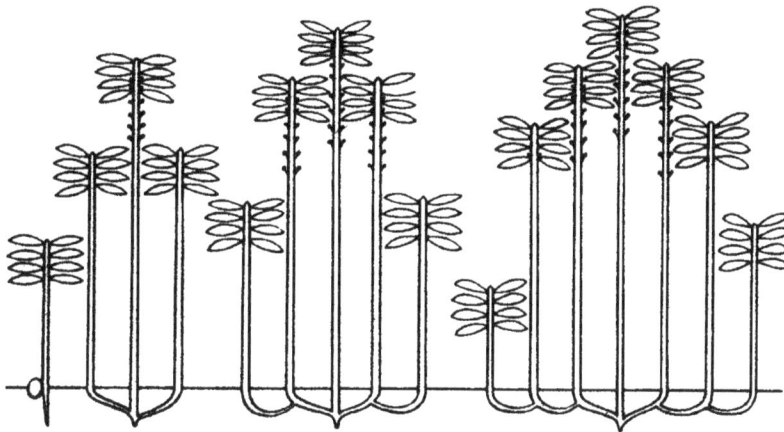

Monocotyledon: *Phoenix dactylifera* (date palm—Palmae).

— Acrotony, i.e., branches not at the base but distal on the axis . . . . 5.

5. Dichotomous branching by equal division of apical meristem . . . .
. . . . **Schoute's model**.

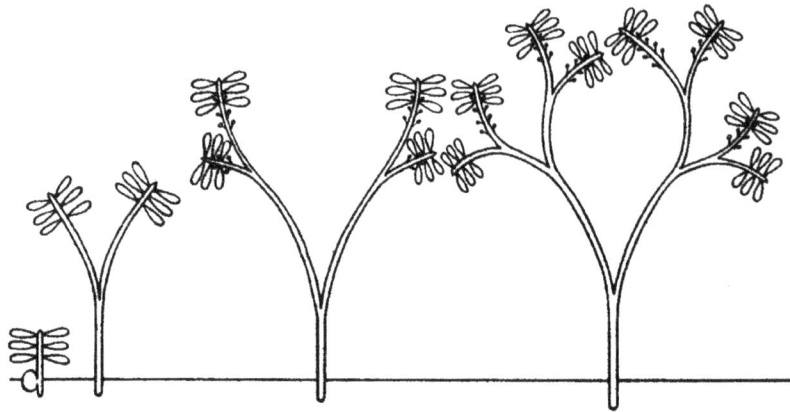

Monocotyledons:

Vegetative axes orthotropic: *Hyphaene thebaica* (doum palm—Palmae).

Vegetative axes plagiotropic: *Nypa fruticans* (nipa palm—Palmae)

— Axillary branching, without dichotomy .... 6.

6. One branch per module only; sympodium one-dimensional, linear, monocaulous, apparently unbranched, modules hapaxanthic, i.e., inflorescences terminal ... . **Chamberlain's model**.

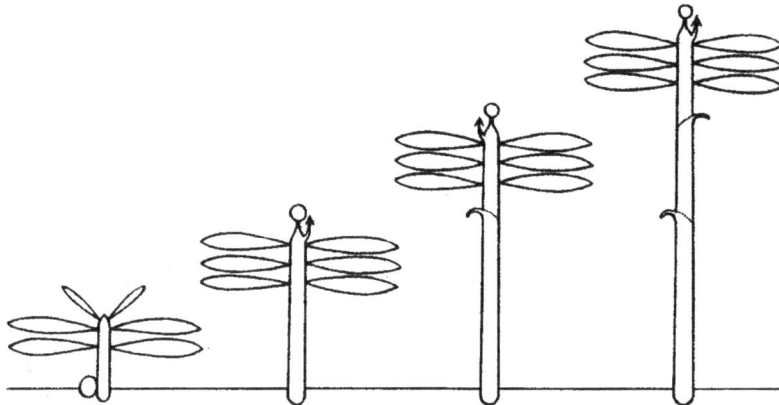

Gymnosperm: Male *Cycas circinalis* (Cycadaceae). Monocotyledon: *Cordyline indivisa* (Agavaceae). Dicotyledon: *Talisia mollis* (Sapindaceae).

— Two or more branches per module; sympodium three-dimensional, nonlinear, clearly branched; inflorescences terminal . . . . **Leeuwenberg's model**.

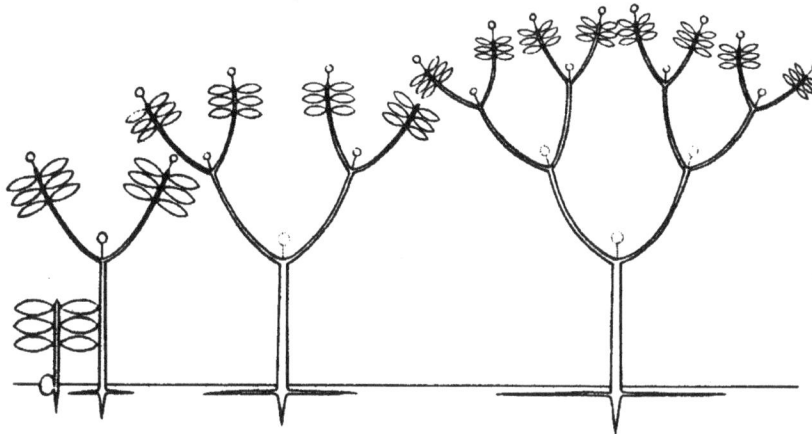

Monocotyledon: *Dracaena draco* (dragon tree—Agavaceae). Dicotyledon: *Ricinus communis* (castor-bean), *Manihot esculenta* (cassava), both Euphorbiaceae.

7 (3). Vegetative axes heterogeneous, i.e., differentiated into orthotropic and plagiotropic axes or complexes of axes . . . . 8.

— Vegetative axes homogeneous, i.e., either all orthotropic or all mixed . . . . . 18.

8. Basitonic (basal) branching producing new (usually subterranean) trunks . . .
. . . . **McClure's model**.

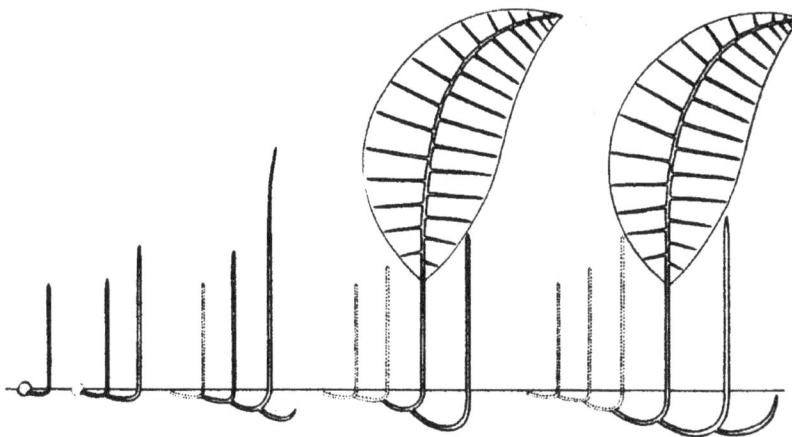

Monocotyledon: *Bambusa arundinacea* (bamboo—Gramineae / Bambusoideae).
Dicotyledon: *Polygonum cuspidatum* (Polygonaceae).

— Acrotonic (distal) branching in trunk formation (never subterranean) ...... 9.

9. Modular construction, at least of plagiotropic branches; modules generally with functional (sometimes with more or less aborted) terminal inflorescences ...
.... 10.

— Construction not modular; inflorescences often lateral but always lacking any influence on main principles of architecture .... 13.

10. Growth in height sympodial, modular .... 11.

— Growth in height monopodial, modular construction restricted to branches ...
.... 12.

11. Modules initially equal, all apparently branches, but later unequal, one becoming a trunk .... **Koriba's model**.

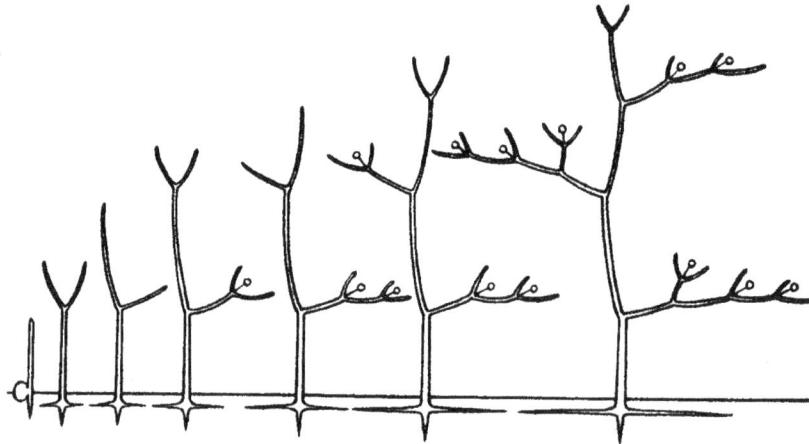

Dicotyledon: *Hura crepitans* (sand-box tree—Euphorbiaceae).

— Modules unequal from the start, trunk module appearing later than branch modules, both quite distinct .... **Prevost's model**.

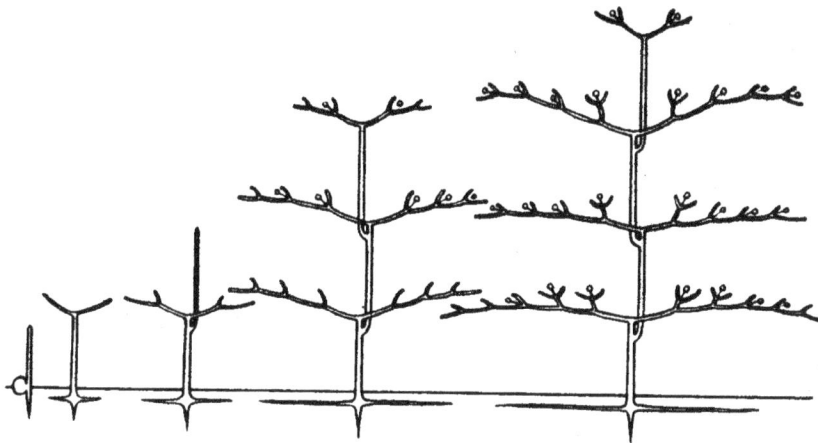

Dicotyledon: *Euphorbia pulcherrima* (poinsettia—Euphorbiaceae), *Alstonia boonei* (emien—Apocynaceae).

12 (10). Monopodial growth in height rhythmic . . . .**Fagerlind's model**.

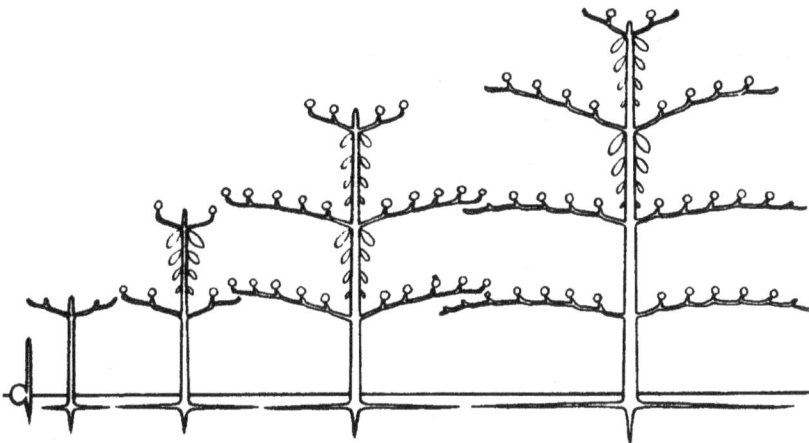

Dicotyledon: *Cornus alternifolius* (dogwood—Cornaceae), *Fagraea crenulata* (Loganiaceae), *Magnolia grandiflora* (Magnoliaceae):

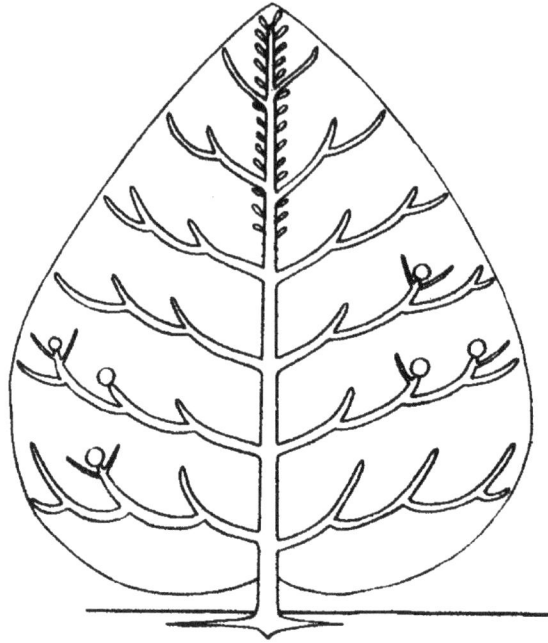

— Monopodial growth in height continuous .... **Petit's model**.

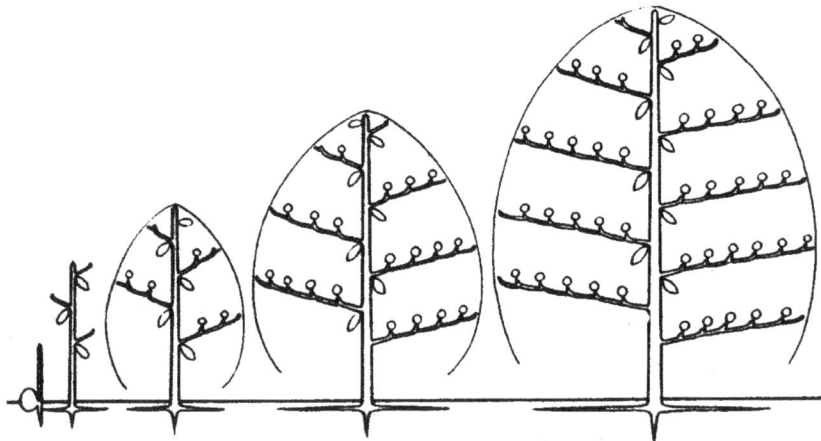

Dicotyledon: *Gossypium* spp. (cottons—Malvaceae).

13 (9). Trunk a sympodium of orthotropic axes (branches either monopodial or sympodial, but never plagiotropic by apposition) .... **Nozeran's model**.

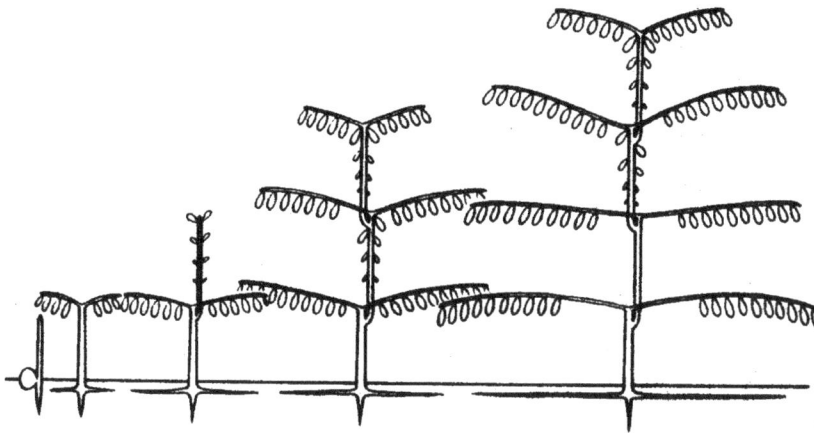

Dicotyledon: *Theobroma cacao* (cocoa—Sterculiaceae).

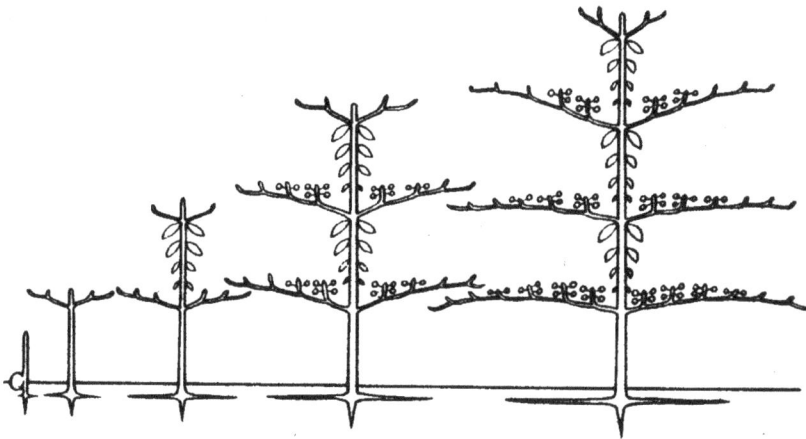

Dicotyledon: *Terminalia catappa* (sea-almond—Combretaceae).

Theoretical Model II defined as an architecture resulting from growth of a meristem producing a sympodial modular trunk, with tiers of branches also modular and plagiotropic by apposition, has still not been recognized in a known example. It would occur here, next to Aubreville's model from which it differs in its sympodial trunk.

— Branches plagiotropic but never by apposition, monopodial or sympodial by substitution .... **Massart's model**.

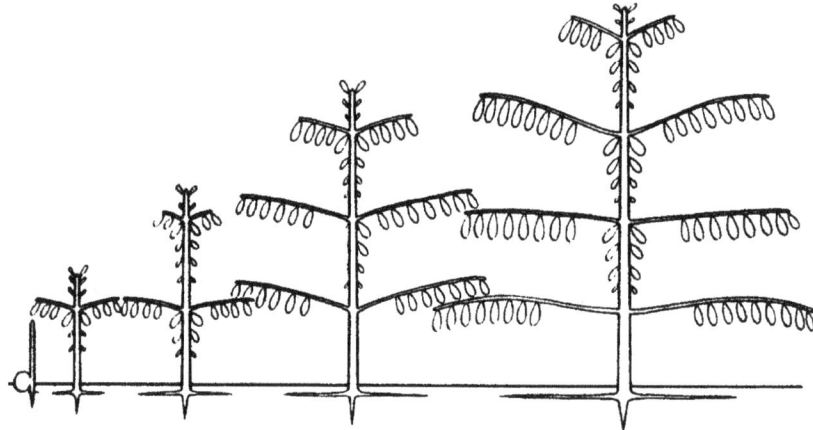

Gymnosperms: *Araucaria heterophylla* (Norfolk Island pine—Araucariaceae). Dicotyledon: *Ceiba pentandra* (kapok—Bombacaceae), *Myristica fragrans* (nutmeg—Myristi◄

16 (14). Branches plagiotropic but never by apposition, monopodial or sympodial by substitution .... 17.

— Branches plagiotropic by apposition .... **Theoretical model I**.

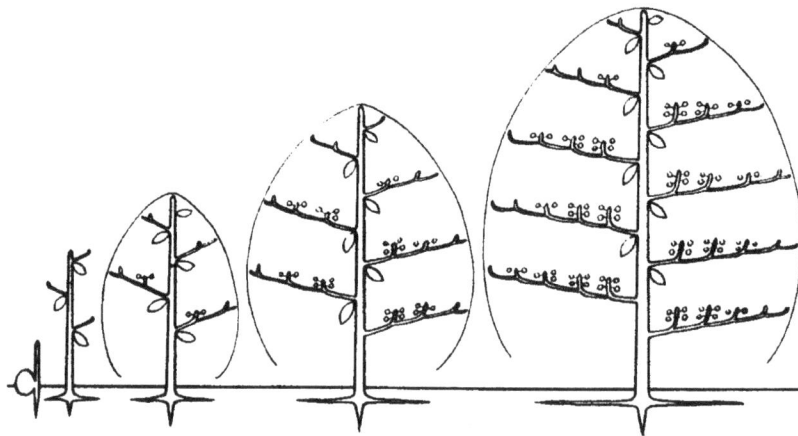

Dicotyledon: *Euphorbia* sp. (Euphorbiaceae)

17. Branches long-lived, not resembling a compound leaf .... **Roux's model**.

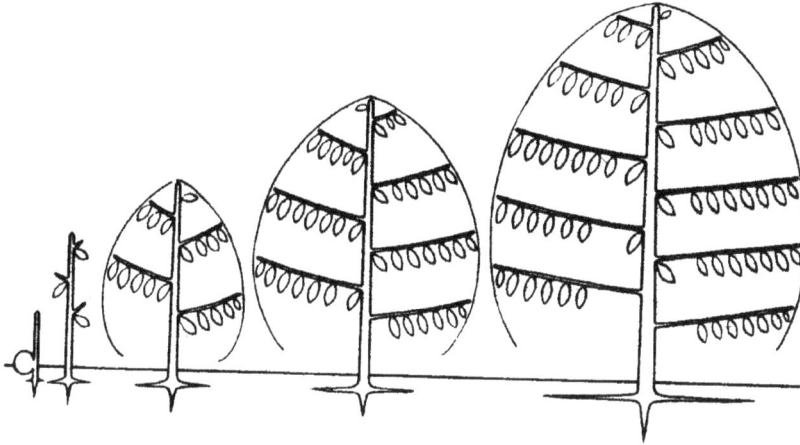

Dicotyledon: *Coffea arabica* (coffee—Rubiaceae), *Bertholletia excelsa* (Brazil nut—Lecythidaceae).

 — Branches short-lived, phyllomorphic, i.e., resembling a compound leaf ......
.... **Cook's model**.

Dicotyledon: *Castilla elastica* (Ceara rubber tree—Moraceae)

18 (7). Vegetative axes all orthotropic .... 19.

 — Vegetative axes all mixed .... 22.

19. Inflorescences terminal, i.e., branches sympodial and, sometimes in the periphery of the crown, apparently modular .... 20.

 — Inflorescences lateral, i.e., branches monopodial .... 21.

20. Trunk with rhythmic growth in height .... **Scarrone's model**.

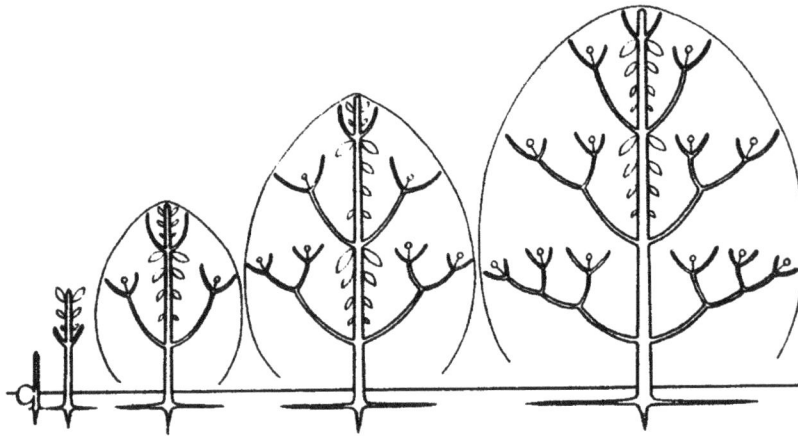

Monocotyledon: *Pandanus vandamii* (Pandanaceae). Dicotyledon: *Mangifera indica* (mango—Anacardiaceae).

— Trunk with continuous growth in height .... **Stone's model**.

Monocotyledon: *Pandanus pulcher* (Pandanaceae). Dicotyledon: *Mikania cordata* (Compositae)

21 (19). Trunk with rhythmic growth in height .... **Rauh's model**.

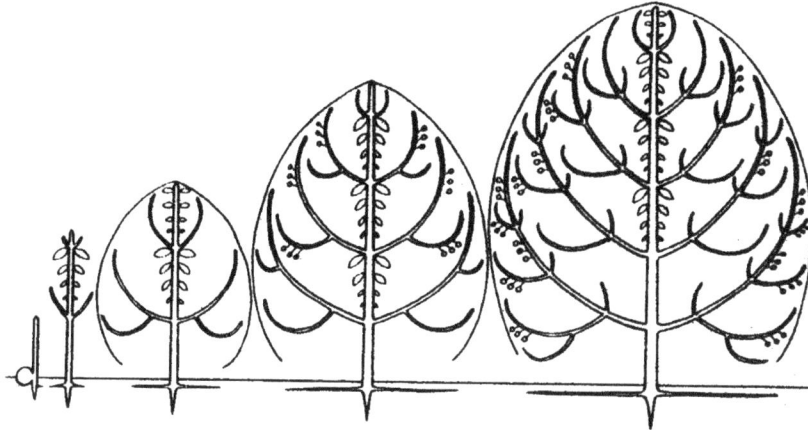

Gymnosperm: *Pinus caribaea* (Honduran pine—Pinaceae). Dicotyledon: *Hevea brasiliensis* (Para rubber tree—Euphorbiaceae).

— Trunk with continuous growth in height .... **Attims'model**.

Dicotyledon: *Rhizophora racemosa* (Rhizophoraceae)

22 (18). Axes clearly mixed by primary growth, at first (proximally) orthotropic, later (distally) plagiotropic .... **Mangenot's model**.

111

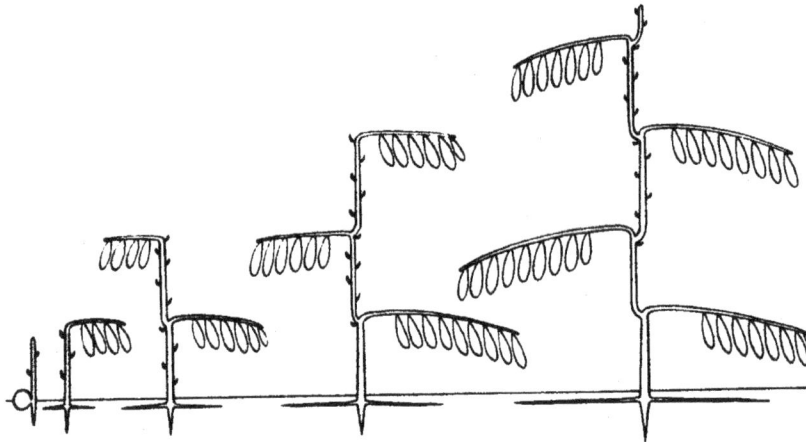

Dicotyledon: *Strychnos variabilis* (Loganiaceae).

— Axes apparently mixed by secondary changes ....22.

23. Axes all orthotropic, secondarily bending (probably by gravity) ....
....**Champagnat's model**.

Dicotyledon: *Bougainvillea glabra* (Nyctaginaceae).

— Axes all plagiotropic, secondarily becoming erect, most often after leaf-fall ...
....**Troll's model**.

Dicotyledon: *Annona muricata* (custard apple—Annonaceae), *Averrhoa carambola* (carambola—Oxalidaceae), *Delonix regia* (poinciana—Leguminosae/Caesalpinioideae)

(a) Trunk a monopodium (e.g., *Cleistopholis patens*—Annonaceae):

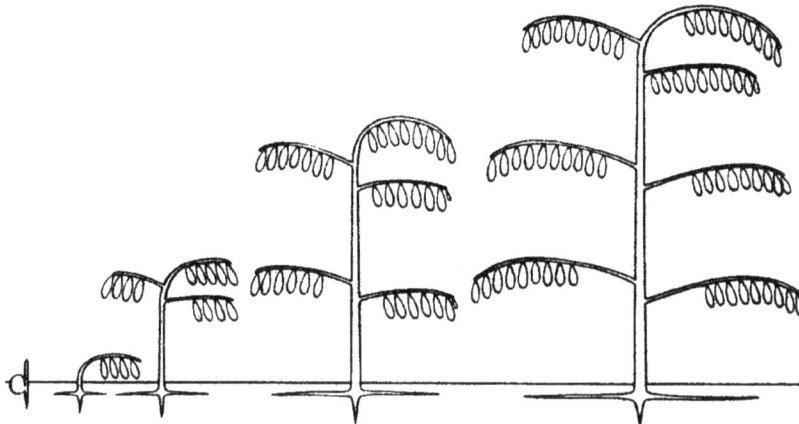

(b) Trunk a sympodium (e.g., *Parinari excelsa*—Rosaceae):

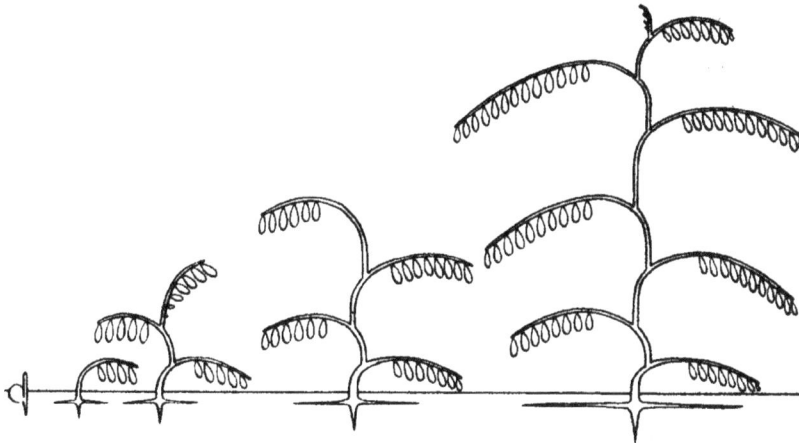

* * *

Some woody vines do not conform with known tree models, e.g. *Triphyophyllum pellalum*, *Ancistrocladus abbreviatus* and *Hedera helix*:

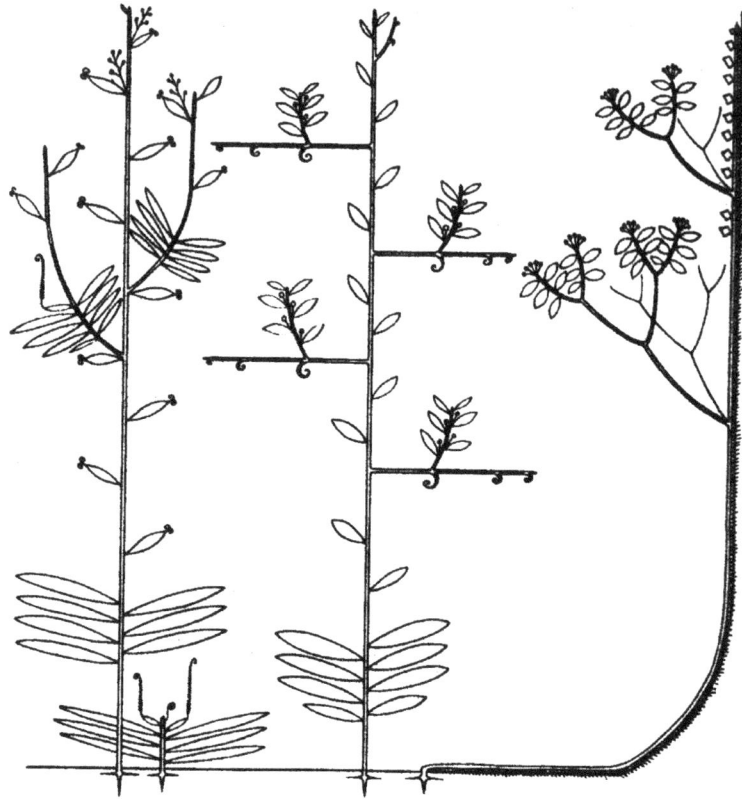